THE LONDON UNDERGROUND
1968–1985

Front cover image
On 4 August 1969, a mixed formation of District Line Q Stock departs from East Putney Station on its Wimbledon branch, bound for Upminster. Of note are the semaphore signal, and the various Southern Railway paraphernalia visible. The station and the branch south to Wimbledon were still owned and managed by British Rail at the time.

Back cover images
Battery locomotives for use on engineers' trains when the current is switched off were first produced to this general design in 1936, and many of them remain in service today. L55, at Neasden Depot on 18 April 1970, was built by British Rail in 1965.

One of the greatest blots on the history of the London Underground is the abandonment of the uncompleted Northern Line extensions after World War Two. Until the end of September 1970, the section between Highgate Depot and the isolated Northern City Line at Drayton Park was used for stock transfers between the two, its only use once British Railways goods trains were withdrawn from the line in 1964. On 4 July 1969, a unit of 1938 Stock is towed through disused Crouch End Station by battery locomotive L21. It is a hot day, and staff riding on it have the cars' doors open on the offside to cool off.

THE LONDON UNDERGROUND 1968–1985

The Greater London Council Years

JIM BLAKE

First published in Great Britain in 2023 by
Pen and Sword Transport
An imprint of
Pen & Sword Books Ltd.
Yorkshire - Philadelphia

Copyright © Jim Blake, 2023

ISBN 978 1 39905 563 5

The right of Jim Blake to be identified as author of this work has been asserted by him in accordance with the Copyright, Designs and Patents Act 1988.

A CIP catalogue record for this book is available from the British Library.

All rights reserved. No part of this book may be reproduced or transmitted in any form or by any means, electronic or mechanical including photocopying, recording or by any information storage and retrieval system, without permission from the Publisher in writing.

Typeset by SJmagic DESIGN SERVICES, India.

Printed and bound in the UK by CPI Group (UK) Ltd., Croydon. CR0 4YY

Pen & Sword Books Ltd incorporates the imprints of Pen & Sword Books Archaeology, Atlas, Aviation, Battleground, Discovery, Family History, History, Maritime, Military, Naval, Politics, Railways, Select, Transport, True Crime, Fiction, Frontline Books, Leo Cooper, Praetorian Press, Seaforth Publishing, Wharncliffe and White Owl.

For a complete list of Pen & Sword titles please contact

PEN & SWORD BOOKS LIMITED
George House, Units 12 & 13, Beevor Street, Off Pontefract Road,
Barnsley, South Yorkshire, S71 1HN, England
E-mail: enquiries@pen-and-sword.co.uk
Website: www.pen-and-sword.co.uk

or

PEN AND SWORD BOOKS
1950 Lawrence Rd, Havertown, PA 19083, USA
E-mail: uspen-and-sword@casematepublishers.com
Website: www.penandswordbooks.com

INTRODUCTION

LONDON'S UNDERGROUND RAILWAYS, with the exception of the short Waterloo & City Line of the Southern Railway (SR), were unified upon the creation of the London Passenger Transport Board (LPTB) in 1933. That authority also took over all bus, tram and trolleybus services within the Greater London area, as well as most bus services outside the metropolis within around 30-40 miles of the centre of London (which became the LPTB's Country Area), together with the Green Line coaches reaching the same area. From 1 January 1948, the whole entity was fully nationalised and renamed the London Transport Executive (LTE), being 'divorced' from British Transport Commission (BTC) control fifteen years later and retitled the London Transport Board (LTB). During those years, the original London Transport 'empire' remained fully intact, but from 1 January 1970, the Country Area buses and Green Line coaches were taken over by National Bus Company subsidiary London Country Bus Services Ltd, with overall control of what was now again called the London Transport Executive passing to the Greater London Council (GLC). That entity in turn had been created in 1965, encompassing the whole of the former London County Council area, most of Middlesex and inner portions of Essex, Kent and Surrey. Its area in fact corresponded almost exactly with London Transport's Central Area, served by its red buses – with the notable exception of the Chigwell and Loughton areas. The new LTE continued to operate its Central Area (red) bus network and the whole of the Underground. However, some Underground lines, of course, extended well beyond the new County of London boundaries, notably the Metropolitan Line to Amersham, Chesham and Watford; the Bakerloo Line to Watford, and the Central Line to Epping and Ongar.

This book looks at the Underground system in the years that the GLC had overall control of London Transport through a variety of photographs I took during those years, until control of London's buses and Underground was snatched away from the GLC by the Thatcher Tory regime in 1984, before it abolished London's elected London-wide government entirely two years later.

The photographs are presented chronologically, and in order to set the scene for the start of the GLC era and to give as comprehensive a picture as possible of the oldest and most interesting rolling stock in use at the time, my survey begins in 1968/69. All of the photographs included are my own, very many of them having never been published before. During my years as a prolific transport photographer (1961-2015), I usually concentrated on rolling stock, or stations and services, due for withdrawal or closure, and this will be evident in many of those presented here. I have not dwelt on the technical aspects of the rolling stock illustrated, as that is not this book's purpose, such details being readily available elsewhere. For readers interested in such things, I would recommend the excellent publications produced by the London Underground Railway Society.

May I thank Colin Clarke for scanning all of my 100,000 or so negatives a few years ago, thereby making it easy for me to locate the pictures included herein, and also John Scott-Morgan of Pen & Sword Books for arranging this book's publication.

JIM BLAKE
Palmers Green
November 2023

GENERAL OVERVIEW OF PASSENGER ROLLING STOCK TYPE ALLOCATION AS AT 1 JANUARY 1970

A. SUB-SURFACE LINES

    Circle Line - CO/CP Stock
    District Line - CP Stock, Q Stock, R Stock
    Metropolitan Line (main section) - A60 & A62 Stock
    Metropolitan Line (East London Line) - Q Stock
    Metropolitan Line (Hammersmith & City Line) – CO/CP Stock

B. TUBE LINES

    Bakerloo Line - 1938 Stock
    Central Line - 1962 Stock (main service): 1960 Stock (Woodford to Hainault shuttle)
    Northern Line - 1938 Stock (including Northern City Line)
    Piccadilly Line - 1938, 1956 Prototype & 1959 Stock
    Victoria Line - 1967 Stock

(N.B.: Some 1938 Stock units included 1949 Stock cars, essentially of the same design. Some Victoria Line 1967 Stock was used on the Central Line's Woodford to Hainault shuttle until the line's extension to Brixton was completed)

***Above and opposite above*: I am** including these two views of the original Highbury & Islington Station on the former Great Northern & City Line, since it was my local station for many years and through whose service I first gained an interest in London's Underground. I took them on 3 April 1968, shortly before its original entrance and lifts were taken out of use, and the station linked with the former North London Railway Highbury Station opposite it at the southern end of Holloway Road. This was in connection with the construction of the new Victoria Line which opened five months later and provided cross-platform interchange with the Northern City Line. Of particular interest is the large LPTB sign at the station's rear entrance in Highbury Place. What price would that fetch today for collectors if it still existed? The original station building still exists, in use as storage and ventilation facilities, and there have been plans in recent years to reinstate it and its lifts to provide step-free access to the tube platforms.

***Opposite below*: Although the** London Underground has a good safety record, it is only natural on such an extensive system that has been in existence for some 160 years that some accidents will have occurred over the years. One such was a derailment, fortunately without casualties, at St. James's Park Station, beneath of all places London Transport's 55 Broadway headquarters on 4 April 1968. LT were well equipped with emergency vehicles to deal with such incidents, and one of them is 1280LD, a specially equipped breakdown tender based on a Leyland Titan PD3A bus chassis. It was one of a number supplied in 1963/64 which replaced earlier vehicles that had been converted from withdrawn buses. It stands outside the station's rear entrance in Petty France. The incident caused total havoc in the evening rush hour, with LT's buses and hired coaches having to provide a replacement service on the Circle and District Lines between Charing Cross (now Embankment) and Sloane Square.

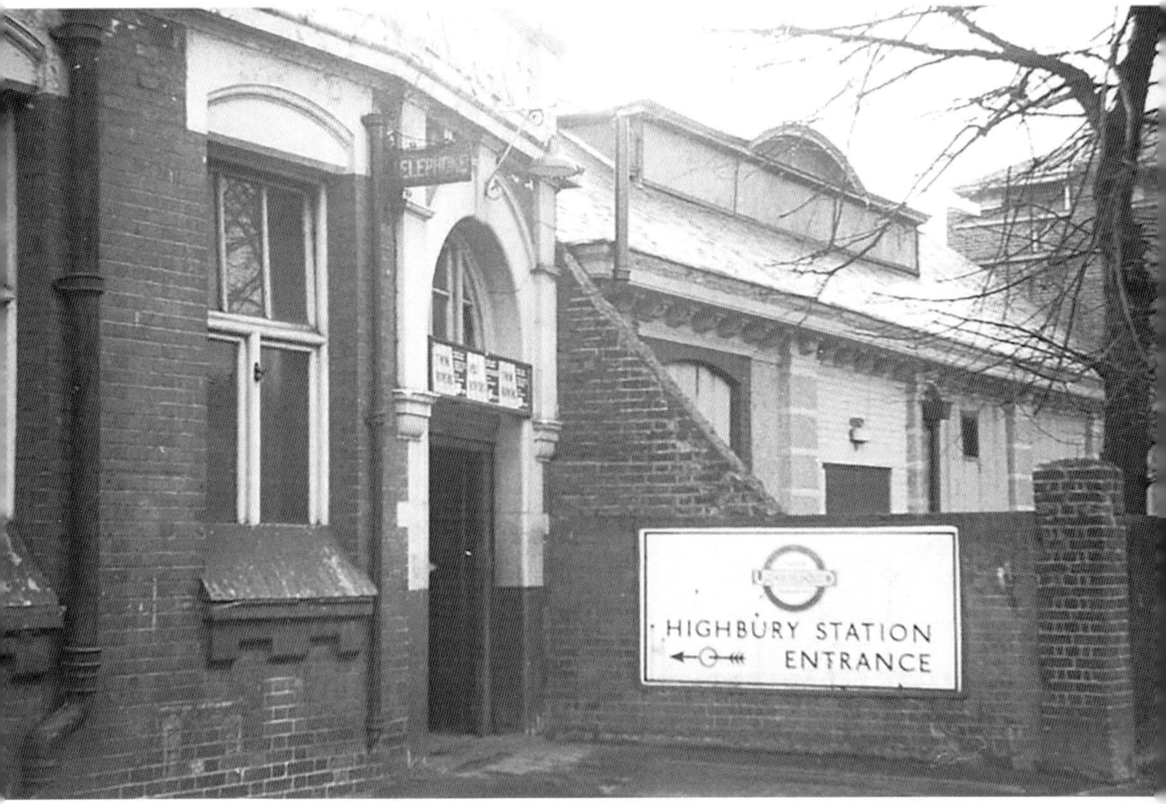
HIGHBURY STATION
ENTRANCE

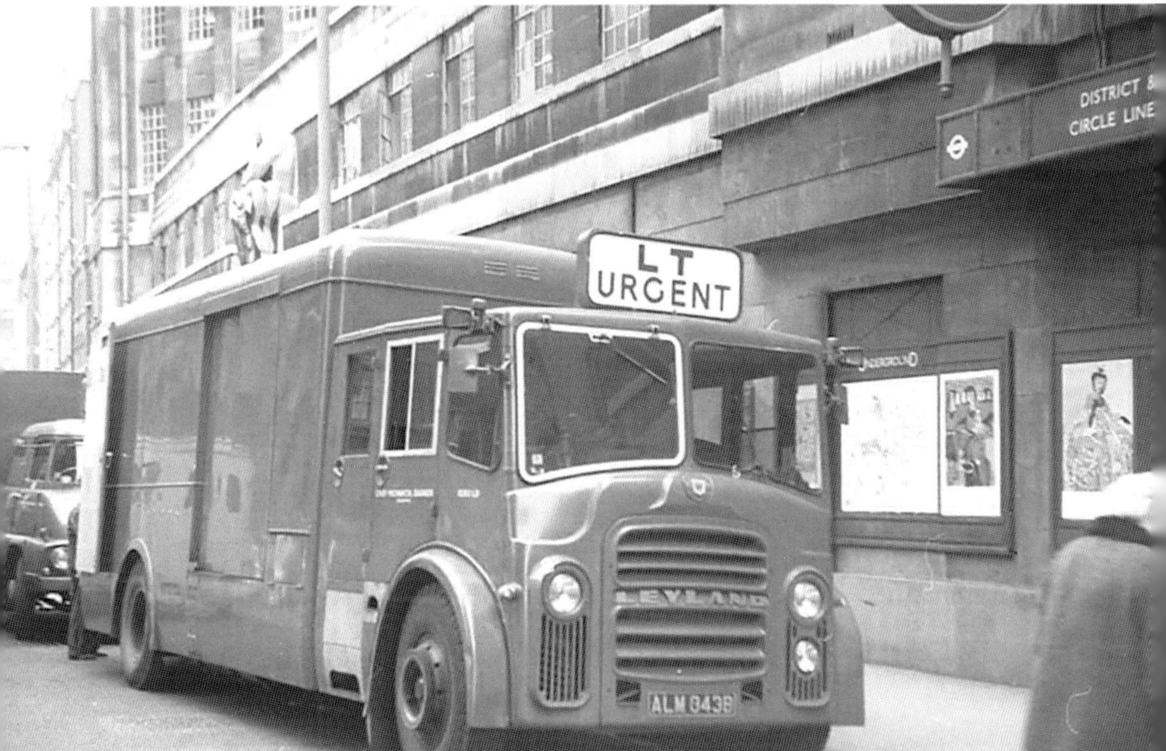
DISTRICT &
CIRCLE LINE
LT
URGENT
LEYLAND
ALM 843B

Above: **One of** the greatest blots on the history of London's Underground is the abandonment of the Northern Line's extensions over the former Great Northern Railway (GNR) branches from Finsbury Park to Highgate, Alexandra Palace and East Finchley, and from Mill Hill East to Edgware. Work on them was very well advanced when it had to be put on hold during the Second World War. Although London Transport fully intended to complete the extensions after the war (and still included them on its Underground maps for several years afterwards), a shortage of manpower, materials and 'above all' funding meant they were abandoned in the early 1950s. Most ironically, at the time of the GLC's assumption of overall control of London Transport, the section between Highgate and Finsbury Park/Drayton Park was still used for transfers of rolling stock to and from the isolated Northern City Line. On a drizzly and miserable 6 May 1969, battery locomotive L21 emerges from Highgate East Tunnel towing a unit of 1938 tube stock to Drayton Park.

Opposite above: **The oldest** passenger rolling stock still in service on the Underground's deep-level tube lines at the time of the GLC's assumption of control of it were a number of Standard (pre-1938) stock trailers formed into units with newer cars. A few survived working with 1960 tube stock on the Central Line's Woodford to Hainault shuttle, but better-known and more numerous were those included in formations of 1938 tube stock on the Bakerloo Line. Because there were fifty-eight of them so utilised initially, they were always referred to as the '58 trailers'. Some had been used on the Northern Line in earlier years. One of them, car 70530, calls at Queen's Park station heading south on 22 May 1969. A British Railways 'sausage' sign for the station is reflected in one of the windows of the car on the left – all four platforms here had such London Midland Region signage at the time.

Opposite below: **On 4 June** 1969, a seven-car formation of Bakerloo Line 1938/1949 tube stock, including one of the '58 trailers', is just about to pass beneath the northbound Metropolitan and British Rail main line tracks north of Harrow-on-the-Hill Station on its way to overhaul at Acton Works. To get there, it will have to continue along the Metropolitan's Uxbridge branch to Rayners Lane, and then reverse to follow the Piccadilly Line to Acton Town. Also of note in this picture is the disused goods shed on the right, in the yard of which one of London Transport's RTW-class buses is being used to teach novice drivers their manoeuvring skills.

Above: **On 30 June** 1969, a Bakerloo Line train similarly formed departs from Queen's Park Station and descends into the tunnels that continue all the way beneath Central London to its Elephant & Castle terminus. An oddity here at Queen's Park is that there are car sheds for the Bakerloo Line at both ends of the station, those at its southern end being visible here above the train's leading end. At this time, most Bakerloo Line trains terminated here, with just a handful continuing, in peak hours only, to Watford Junction. British Railways London Midland Region local dc electric trains to that point shared their tracks with the Bakerloo Line not only on the service to and from Euston, which still runs today, but also with services to and from Broad Street.

Opposite above: **The oldest** passenger stock of all on the Underground at this period operated on the District and East London Lines. These were classified Q Stock by London Transport, though were produced by a number of manufacturers for the District Railway, which had its own classifications for each variety. On 1 July 1969, Q27 driving motor car 4253 leads an eastbound District Line train entering Bromley-by-Bow Station bound for Upminster. This had been one of the District's K class, and as the '27' in its new classification suggests, the class dated from 1927. All of the Q Stock either inherited from the District Railway or, as in the case of the Q35 Stock ordered by it but delivered after London Transport's creation, perpetuated the clerestory roof design dating back to the very earliest electric stock.

Opposite below: **As the** set number 121 on the trailing driving motor car of this formation shows, it is the same train as that in the previous picture, now departing from Bromley-by-Bow. However, this car, No.4402, is of drastically different design to the other clerestory-roofed Q Stock in the train. It is a Q38, delivered between 1938 and 1941 along with the similarly-bodied O and P stocks which worked the Circle and Metropolitan Lines. Somewhat oddly, both Q38 driving motor cars and trailers were mixed indiscriminately with the older stock, often giving a somewhat eccentric appearance to their trains. In this one, however, only car 4402 is a Q38, all of the others are earlier stock.

THE LONDON UNDERGROUND, 1968–1985 • **13**

Above: **Also at** Bromley-by-Bow that day, CO Stock driving motor car 53055 is at the trailing end of a Hammersmith & City Line train. Although the bodywork is the same design as the Q38 in the previous picture, a brief comparison of the two cars will show slight differences in the appearance of their cab ends. Most notable is the difference between their coupling mechanisms – that on Q38s had to be compatible with the older District Q Stock. Ten years or so previously, changes were made to the O and P Stocks' control equipment, leading to their being reclassified CO and CP. At the time I took this picture, the Hammersmith & City Line, still officially part of the Metropolitan Line, only ventured east of Whitechapel to Barking in Monday to Friday peak hours.

Opposite above: **The very** oldest operational rolling stock of all on the Underground at this time was the Q23 Stock, originally District Railway class G. Dating from 1923, their cab front looked much more antiquated than the Q27s produced only four years later. Their destination boards were carried externally, rather than behind a glass panel too. Framed by the ornate ironwork of the London, Tilbury & Southend Railway (LT&SR) footbridge at the London end of Bromley-by-Bow Station, driving motor car No.4188 leads an Ealing Broadway-bound train.

Opposite below: **At this** time and until only a few years ago, Whitechapel was a six-platform station, with its two outer platforms in the upper part normally used for terminating Hammersmith & City trains in off-peak times and at weekends. It is the first open-air station on the District Line's eastern section, though eastbound trains descend into the tunnel again after leaving it. Also on 1 July 1969, Q27 driving motor car 4252 is at the trailing end of an Upminster-bound train doing so. A Q38 car is visible in its formation at the extreme right of the picture. Owing to the construction of Crossrail, the two centre tracks here were removed to provide extra passenger circulating space. The two outer tracks thus became the through tracks, no longer needed for terminating Hammersmith trains, since these had long since been extended daily to Barking. Below this area are the two tracks of the East London Line, at the eastern end of the station.

THE LONDON UNDERGROUND, 1968–1985 • **15**

Above: **At Whitechapel,** the East London Line passes beneath the District Line tracks as noted above. A curious state of affairs existed for many years, whereby although this line was officially part of the Metropolitan Line complex, rolling stock was provided by the District Line. Q Stock had replaced the earlier ex-District F Stock in 1963, and here Q27 driving motor car 4321 leads a four-car formation comprised of two clerestory-roofed cars and two Q38s. It arrives heading south from Shoreditch, then served only in peak hours, and on Sunday mornings for the benefit of people attending Club Row Sunday market with its famous beigel shop and bakery. Typifying somewhat sloppy destination displays on this short line at this time, it does not notify intending passengers whether it is bound for New Cross or New Cross Gate, displaying only 'Metropolitan'. Presumably, the train describers on the platform will put intending passengers right.

Opposite above: **There is** nothing to indicate that the northbound train headed by Q23 driving motor car 4184 is bound for Shoreditch, either, as it offers an interesting comparison between its own antiquated-looking cab area and the slightly more modern lines of Q27 driving motor car 4362 at the trailing end of a southbound train on the adjacent platform. At this time, the East London Line platforms at Whitechapel were somewhat gloomy and forbidding, apart from the area immediately adjacent to their entrances and connections with the 'main' lines above. This perhaps befitted the place, since just eighty-one years previously, Jack The Ripper had committed his first ritual murder in Buck's Row just behind the station, and when I took this picture, some very elderly people who remembered his reign of terror still lived locally and may have been related to him even.

Opposite below: **Many Q38** Stock trailers were converted after the war to become driving motor cars, forming the basis of the new R Stock for the District Line. They were classified R38, and eventually marshalled into formations including newly-built cars of basically the same design, of types R47, R49 and R59. The older cars were originally red, but the later new ones were delivered in unpainted aluminium. By the late 1960s, all of those that had been red had been painted silver to match them. On 4 July 1969, a typical train of R Stock approaches Barking Station, working a Richmond-bound service. It is about to burrow beneath the British Rail Tilbury-line tracks, segregation of these and the LT tracks here having been made some ten years previously as a prelude to the overhead electrification of the former LT&SR lines.

THE LONDON UNDERGROUND, 1968–1985 • **17**

Above: **On the** same occasion, R38 driving motor car 21128 is at the trailing end of a District Line service terminating at Dagenham East. In the distance beyond it may be seen Barking sidings where some District Line trains were stabled at weekends. On the left, the large station car park occupies the site of the former goods yard.

Opposite above: **Two trains** of District Line Q Stock are stabled in Barking sidings on Saturday 4 July 1969. Nearest the camera is Q27 driving motor car No.4370, whose train includes two Q38 cars. It is hard to believe that these are only two or three years newer than some of the last clerestory-roofed Q Stock. Their modern design epitomises the progressive nature of London Transport in the late 1930s.

Opposite below: **Another train** of R38 Stock is visible in this view of Lillie Bridge Depot and its surroundings, approaching West Kensington Station on 26 July 1969. The electrified track in the foreground leads to the eastern end of this station. At this period, the depot was used only for non-passenger rolling stock operating engineers' trains and so on, and two of the typical Underground battery locomotives, L23 and L27, stand outside one of the sheds. To the left of them is a Standard Stock driving motor car, one of many converted to depot shunting locomotives, termed pilot motor cars, or for hauling ballast trains. Two steam engines were also still based here at this time and will be illustrated later. On the extreme left of the picture is the British Rail West London Line, at the time used mainly for cross-London freight services and occasional inter-regional passenger trains, together with rush hour passenger trains between Clapham Junction and Olympia. It would not regain a comprehensive local passenger service until the 1990s, but today has both a busy London Overground service and also trains linking places like Harrow and Watford with the former Southern Region. The large building in the top left-hand corner of the picture is Earls Court exhibition hall, alas recently demolished.

Above: **On 1 August** 1969, CP Stock driving motor car 53223 is at the trailing end of an 'outer rail' (i.e., clockwise) Circle Line train departing from Gloucester Road Station. Until replaced by new A Stock in the early 1960s, most CP Stock trains worked on Metropolitan Line Uxbridge services. Now, they were due to be replaced by new C69 Stock units on Circle and Hammersmith & City Line services, though they were moved to the District Line for further use, replacing older Q Stock.

Opposite above: **Also at** Gloucester Road that day, District Line set No.113 arrives on a Wimbledon-bound service, led by Q27 Stock driving motor car 4364. Although what is visible of the train here shows all clerestory-roofed cars, there will probably be a Q38 car further back in the formation.

Opposite below: **Those not** familiar with the Underground will probably be surprised to see that steam engines still survived on it at this period. They were used for engineers' trains and, especially, hauling waste material from throughout the sub-surface lines to Watford tip sidings, near the junction between the Metropolitan main line to Amersham and its Watford branch. Retained especially for them was a water tower at the terminus of that branch, at which loco L90 fills up on 4 August 1969, before running around its train and returning south to Neasden Depot. This is one of a number of ex-Great Western Railway 5700 class 0-6-0 pannier tanks which replaced London Transport's indigenous steam engines between 1956 and 1963, having originally been G.W.R. No.7760 built in 1930. A train of Metropolitan Line A Stock, a type which would see more than fifty years' service, completes the picture.

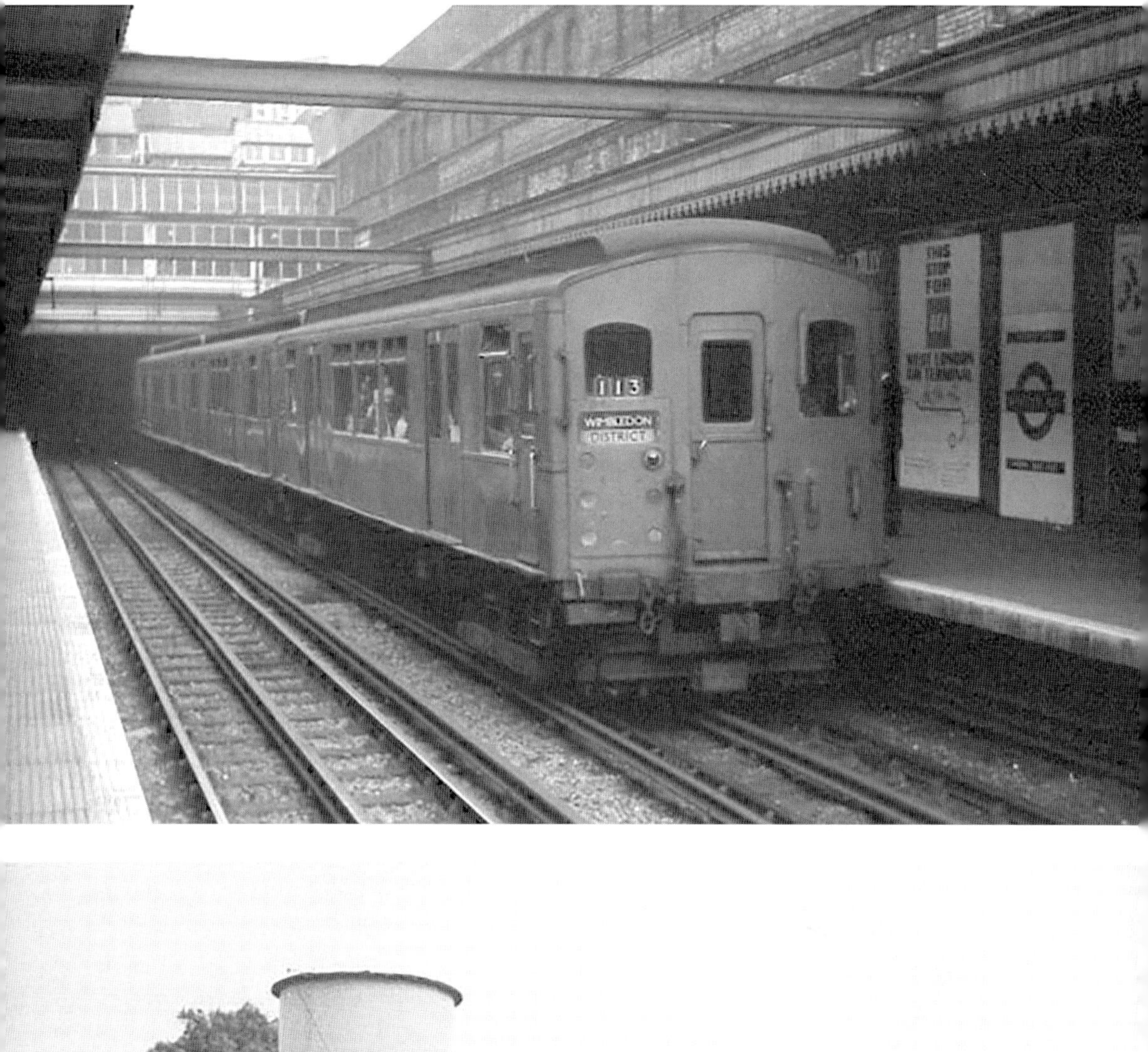
113
WIMBLEDON
DISTRICT

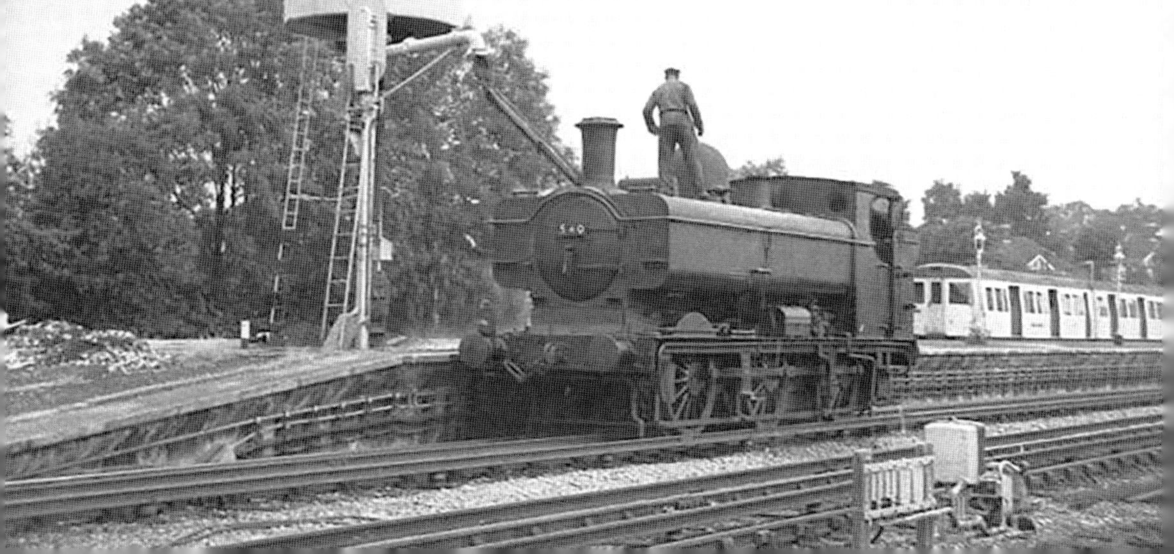

Although the Piccadilly Line's Standard Stock was replaced by new 1959 Stock units in the early 1960s, fifteen units of 1938 Stock were in use on the line for many years, lasting until the 1959 Stock trains were replaced by new 1973 Stock. Also on 4 August 1969, driving motor car 11030 has just come out of the siding at Rayners Lane to begin its long journey through and beneath London to distant Cockfosters.

I rode on the 1938 Stock train illustrated above to Acton Town, where by lucky coincidence the Underground's track testing unit put in an appearance. This was formed of two driving motor cars of 1960 Stock along with two Standard Stock trailers. Car No.3910, one of the driving motors, is nearest the camera. Something of a resemblance to the Victoria Line's 1967 Stock is evident. The unit still exists today, though the two trailers have been replaced by a 1973 Stock car.

My next port of call that day was East Putney, where a train of District Line CP Stock bound for Wimbledon has just arrived. Driving motor car 54221 is in the lead. On the right are the disused platforms on the former SR link to the Reading and Windsor lines just east of Putney Station. This was, and still is, used for empty stock movements to and from the British Rail depot at Durnsford Road, Wimbledon.

On the same occasion, the next train bound for Wimbledon is formed of Q Stock, with Q27 car 4364 leading. At first sight, it appears to be formed entirely of clerestory roofed stock, but a closer look reveals a Q38 car towards the rear of the train.

Led by a Q27 driving motor car, an eastbound District Line train departs from East Putney Station, under the command of a British Rail semaphore signal. At this period, the Wimbledon branch of the District Line south of Putney Bridge Station was still owned by British Rail's Southern Region. Also of note in this picture are the passenger door control buttons on the Q Stock cars, meant for use at open-air stations. They had, however, been taken out of use at this time but the facility would be reintroduced on the District's new D Stock some ten years later.

Another Q27 driving motor car, 4293, is at the trailing end of the same Upminster-bound train. Other Southern Railway relics are visible in this view, the typical SR-type lamp post and the affixed car-stop sign on the left, and the permanent way hut on the right. The latter was in SR green and cream colours, too, as were the platform buildings. The station is in marked contrast with the new buildings on the right.

The northernmost station on the District Line's Wimbledon branch is West Brompton, where another Q27 driving motor car, 4366, is at the trailing end of this train departing for Upminster. At this period, West Brompton was closed on Sundays. In more recent times, the then-disused adjacent platforms on the West London Line have been reopened for London Overground and Southern services, making the station much busier than it was in 1969 and of course open daily. The vast bulk of the Earls Court exhibition centre is also visible in this picture, above the station buildings.

The next eastbound District Line train in the evening rush hour of 4 August 1969 to call at West Brompton is led by Q38 driving motor car 4413. The disfigured hut on the westbound platform illustrates how even back in 1969, the plague of graffiti sometimes afflicted the Underground.

Above: Also in the evening rush hour of 4 August 1969, 1938 Stock driving motor car 10029 is at the head of a westbound Piccadilly Line train bound for Rayners Lane. It speeds non-stop through Stamford Brook Station, at the time served only by District Line trains. The architecture of the eastbound District Line platform on the left is typical of the time when (in 1932) the Piccadilly Line was extended westwards from Hammersmith.

Opposite above: On the same occasion, driving motor car No.1000, numerically the first of such cars amongst the 1956 Stock prototypes, is at the trailing end of an eastbound Piccadilly Line service heading for Cockfosters through Stamford Brook Station. At first sight identical to a 1959 Stock car, it may immediately be distinguished from one by retaining the old-style lights on the cab front, which was used to indicate each train's destination. A comparison with the 1938 Stock unit in the previous picture immediately shows how the design of the 1956 prototypes, and the production batches of 1959 and 1962 Stock which followed them, was derived from that of 1938 Stock. Meanwhile, it is also interesting to compare this with that of the eastbound District Line R Stock train which is calling at the station. The car of this on the left is one of the post-war built non-driving motor cars of type R47, R49 or R59.

Opposite below: Q Stock was still quite plentiful on the District Line at this time but would soon be relegated to rush hour use only, and then finally banished to the East London Line 'backwater'. Q27 driving motor car 4199 heads an eastbound train for Upminster arriving at Stamford Brook working set No. 056.

Above: **At the** trailing end of the same formation is Q23 Stock driving motor car 4184. To illustrate how this is truly a mixed formation, the third car along is a Q38 which is just visible on the extreme right of the picture.

Opposite above: **The following** day, 5 August 1969, I again captured on film one of the (usually) weekly stock transfers between the main Northern Line's Highgate Depot and the isolated Northern City Line at Drayton Park. These used the former GNR/London and North Eastern Railway (LNER) branch from Highgate to Finsbury Park, via Crouch End and Stroud Green, which should have become part of the Northern Line complex in 1940/41. Towed by the usual battery locomotive employed on such missions, L21, a four-car unit of 1938 Stock with driving motor car 11252 at the rear crosses the East Coast Main Line just north of Finsbury Park Station, where the branch joins the main line. Although the steam-operated services on the branch which the Northern Line was to replace (and actually did of course, north of East Finchley) joined the main line either side of the through tracks at Finsbury Park Station, tube services in *both* directions would have called on the eastern (up/southbound) side of the station, all thereby crossing the bridge diagonally above the through main line tracks. The need to provide better headroom for the main line's 25kv overhead electrification meant the bridge had to be either rebuilt or removed – sadly, the latter took place as it was impracticable to rebuild it solely to cater for these stock transfers. This therefore spelt final doom for the branch.

Below: **This view** of car 11252, awaiting the train's path through Finsbury Park Station and then on to the branch heading for Canonbury, Dalston Junction and Broad Street and down the ramp into Drayton Park Station, shows in the foreground some of the cable-runs built for the abandoned extension. Such equipment, and most of the cabling and conductor rails for the electrification, was delivered to site with much actually installed when completion of the extension was put 'on hold' during the war. Indeed, both this and the previous picture could for all the world show a tube train actually in service on the line, since the battery locomotive is only remotely discernible in the previous picture. Moreover, much of the 1938 Stock intended for the extension was actually the property of the LNER until both that entity and London Transport were nationalised at the beginning of 1948. This actual car was one such example, and many of them still carried plates on their solebars reading 'Property of LNER' until the end of their Underground service lives in the 1970s/80s.

Above: **More mundane** when I photographed it on 6 August 1969, CO Stock driving motor car 54038 arrives at Moorgate Station on a clockwise ('outer rail') Circle Line service. On the left is an eight-car train of Metropolitan Line A Stock stabled in one of the terminal platforms here, as they were during off-peak periods and at weekends at this time. This sub-surface part of Moorgate Station was badly damaged during the blitz, particularly in the 'second great fire of London' on 29 December 1940, and never fully rebuilt. Now, however, it is being encroached upon by the grotesque Barbican development (also, finally, building over what had for nearly thirty years been wartime bomb sites nearby), which will eventually roof over the station completely.

Opposite above and opposite below: **An oddity** for many years on the Bakerloo and latterly the Northern Line was 1938 Stock driving motor car No. 10306, also one of the 'LNER' cars. In the early post-war years, it was rebuilt with some of its side windows, including those on the passenger doors, extended into the roof. This was done in order to provide standing passengers with better outward vision, to avoid them missing their stations, and as a possible prototype for future tube stock. Although the subsequent 1959 and 1962 Stock cars did not have this feature, 1967 Stock onwards did perpetuate it for their doors. These two views find 10306 awaiting departure from Edgware with a service to Kennington via Charing Cross on 7 August 1969. Also of interest in these views is the 'unfinished' appearance of the station roof, the left-hand supports of which appear to have fittings for a further structure. Presumably this was for the extra platforms intended for the extension to Bushey Heath, a fair amount of work on which (not least the car depot and maintenance facilities at Aldenham) had been done when it had to be placed in abeyance during the war. As is well known, this part of the uncompleted Northern Line extensions too was abandoned in the early 1950s. A third terminal platform, on the left, did come into use.

THE LONDON UNDERGROUND, 1968–1985 • **31**

Above: **The other** 'missing link' in the uncompleted Northern Line extensions was, of course, the section between Mill Hill East and Edgware. This single-track ex-GNR branch of the LNER from Finchley, Church End to Edgware was to be doubled when electrified and was actually closed to traffic a week after the Second World War began, in order to facilitate work, with a substitute bus service being provided until its completion. All required new over- and under-bridges were installed and this view of 1938 Stock driving motor car 10216 leading a seven-car train arriving at Mill Hill East, also on 7 August 1969, clearly shows one of these at the station's London end, along with the space where the second track should be. The second track, along with conductor rails, was installed for most of the way, but later removed for use elsewhere.

Opposite above: **The short** stretch of this branch between Finchley Central (LT's new name for Church End) and Mill Hill East was electrified and brought into use in May 1941, largely owing to the presence of a military barracks nearby. But that is as far as it got. The original GNR station never gained its second platform and remains largely in original condition to this day. This view, also on 7 August 1969, shows the end of the line at Mill Hill East. Conductor rails had been installed beyond this point towards Edgware and would remain unused as they also did between Finsbury Park and Alexandra Palace. Trains would have continued beyond Edgware (where the line was to be diverted into the existing Northern Line station on the branch from Golders Green) to serve the new extension to Bushey Heath, rather than its service being extended from the existing branch. LNER and then British Railways Eastern Region goods trains continued to serve the old GNR Edgware terminus until May 1964, then part of the line's alignment at Page Street, Mill Hill was needed for a slip road serving the M1 motorway's new southward extension. Tracks were then soon lifted, and recent residential development across much of the former trackbed would seem to have killed the railway link between Mill Hill East and Edgware once and for all.

Below: **Making a** mockery of the excuse of London's green belt being used to 'justify' the abandonment of the Finsbury Park to Alexandra Palace and Mill Hill East to Edgware sections of the uncompleted Northern Line extensions when the former was completely built-up before the war, and the latter almost so (as it is today), extensions to the Central Line which were also put on hold during the war were all completed after it, except for a short stretch between West Ruislip and Denham. In the East, the line penetrated deep into rural Essex, Epping being reached in 1949 and, finally, the entirely rural single-track section beyond to Ongar was electrified in 1957. Always operated as a shuttle between Epping and Ongar (except for workings to or from the latter point and the trains' depot), it was never to fulfil the expectations of the 1935-1940 New Works Programme within which it was built, owing of course to the green belt. Driving motor car 1504 of one of the Central Line's fleet of four-car 1962 Stock units, operating on its own rather than in a formation of two of them, as usual for this lightly used section, hopefully awaits passengers at its Ongar terminus also on 7 August 1969.

Above: **The least-used** station on the entire Underground system at this time was Blake Hall, on the Epping to Ongar section, which seldom could count its passengers each day in more than single figures. Being situated in the middle of nowhere, and even some distance from the establishment from which it took its name, did not help. Moreover, a decent London Transport Country Area bus route, the 339, covered the same route between Epping and Ongar. In this view on 7 August 1969, the gentleman emerging onto its single platform is not a passenger, but the station manager, who also lived in the adjoining station house. Perhaps he is surprised to see me, a stranger, alighting there – but I was not alighting, only taking this photograph from car 1504's doors.

Opposite above: **Following the** successful use of the 1956 Stock prototype units on the Piccadilly Line, similar units classified 1959 Stock were ordered to replace all that line's Standard Stock. On the Central Line, however, a different approach was taken in that new driving motor cars termed 1960 Stock were ordered, but their trailers were provided by refurbishing Standard Stock cars. However, after only a few of these hybrid units had been completed, London Transport decided not to proceed further with this project, partly because the Central Line's ageing Standard Stock compared unfavourably with the new 25kv overhead electric units introduced in November 1960 on many of the suburban services into and out of Liverpool Street, whose commuters would notice the striking difference between main line and tube stock, and partly because the conversion of the older cars to run with the new 1960 Stock was proving more costly than anticipated. Therefore, deliveries of new 1959 Stock were diverted to the Central Line from the Piccadilly Line until further cars of the same type, termed 1962 Stock, could complete the job of ousting Standard Stock. Once these had become established, the 1960 Stock trains were banished to the lightly-used Hainault to Woodford shuttle. Also on 7 August 1969, driving motor car 3905 heads a four-car unit into Woodford Station.

Opposite below: **The second** car in the formation, 4905, is one of the earlier Standard Stock trailers converted to run with the new 1960 Stock driving motor cars. They were painted silver to match these unpainted aluminium cars. In common with 1956, 1959 and 1962 Stock formations, only the driving motor cars carried their London Transport fleet names, which were in red and underlined in traditional style.

Above: **Unlike its** partner in this formation, Standard Stock trailer No. 4904 is one of the later deliveries and as well as having additional single doors at each end, which 4905 does not, has a more modern roof styling. Eventually, surviving 1960 Stock units had these trailers replaced by 1938 Stock ones and, somewhat ironically, the trains were painted red to match them. They ended up working the Epping to Ongar shuttle service. Interestingly, the new 1960 Stock cars were built by Cravens of Sheffield, whereas the 1959 and 1962 Stock was built mainly by Metro-Cammell in Birmingham, although the non-driving motor cars needed for the Central Line's eight-car formations were built by British Railways' Derby Works and at first inserted into the seven-car formations borrowed from the Piccadilly Line. Cravens were more than compensated for their loss of the large order of new cars for the Central Line by getting the orders for the Metropolitan Line's A60 and A62 Stock four-car sets, some of which lasted until as late as 2012, seeing more than fifty years' service.

Opposite above: **I rode** on the unit illustrated here, which has 1960 Stock driving motor car 3904 leading when arrived at Hainault, around the still rural northern section of the Hainault loop from Woodford. Of note in this picture is how the platform buildings in the foreground are typical of the 1935-1940 New Works Programme of which the Central Line's extensions were part, whereas those on the opposite platform (as well as those at Woodford) are still the Great Eastern originals.

Opposite below: **Another 'backwater'** of the Underground at this period was the East London Line, particularly its northern terminus at Shoreditch which was actually very close to the City and the busiest lines in Central London. At the start of the evening rush hour of 7 August 1969, Q27 driving motor car 4311 is at the head of a four-car unit waiting hopefully for passengers there when trains have just begun to serve the station. Its second platform on the right has long been taken out of use. Originally trains had continued from here into Liverpool Street terminus, and as recently as the early 1960s the link was still used for British Railways goods trains from the Eastern Region through the Thames Tunnel to the Southern Region, and even seaside excursions. Above the station on the right is the former Great Eastern Railway Bishopsgate Goods Station, destroyed by a disastrous fire in 1964, and left derelict for many years afterwards.

Above: **A ride** on the unit illustrated above took me through Brunel's famous Thames Tunnel between Wapping and Rotherhithe to New Cross Gate, where fellow Q27 driving motor car 4299 heads the next East London Line service to arrive. At this period, extensive Southern Region sidings still existed here, upon which a British Railways standard Mk1 carriage completes the picture, despite all regular main line services here being operated by multiple units at the time.

Opposite above: **Q38 driving** motor car 4420 is at the other end of the newly arrived train, whose guard is just about to open its doors and is visible at the far end of this car. In contrast to the Q27 at the other end which showed its destination correctly, no display other than 'Metropolitan' is shown on it. The track on which the train has terminated was still a through track that continued beyond the station to join Southern Region metals. In 2010, of course, the through link was restored when London Overground services commenced, taking over the East London Line and continuing to West Croydon and Crystal Palace from this station.

Opposite below: **In contrast,** the East London Line's other southern terminus, at New Cross, ended in a dead end stub on the east side of the Southern Region main line through the station, as it still does today. On the same evening as the previous picture, another Q38 Stock driving motor car, 4408, awaits departure for (presumably) Shoreditch. For once, it is coupled to a Q38 trailer, though the unit's remaining cars, out of the picture, are earlier clerestory-roofed Q Stock. I always found it odd that Q38 cars were not formed into their own units for the East London Line; evidently this was owing to an imbalance between motor and trailer cars of the type, and perhaps it was easier to mix all the different varieties of Q Stock up anyway.

THE LONDON UNDERGROUND, 1968–1985 • **39**

Above: **I spent** the rest of the evening rush hour of 7 August 1969 recording scenes of Q Stock on the District Line's eastern section. Driving motor car 4374, another Q27, brings an Ealing Broadway-bound train into Plaistow Station. The platforms on the main line tracks on the right have long since been taken out of use.

Opposite above: **Also at** Plaistow that evening, Q23 Stock driving motor car 4248 is at the trailing end of an Upminster-bound service, as one of the Eastern Region electric multiple units built for the electrification of the LT&SR line services at the start of the decade speeds through the station on its way to Fenchurch Street. Of note is the British Railways (Eastern Region, blue coloured) sign on the right – all of the District Line stations out to Upminster were still managed by that authority at the time.

Opposite below: **Further to** the East that evening, at Becontree Station, another eastbound District Line train illustrates the complexity of the Q Stock formations. With a Q23 Stock car at the rear, the next car, 08808, is a former Q31 (originally class L) driving motor car that has been converted to a trailer. Its driving cab door at the far end is still in place, though sealed up, and seats occupy where the driver's controls, etc. once were. Many such cars were similarly converted after the war, to replace Q38 trailers which had been appropriated for conversion into R38 driving motor cars. The third car along in this formation is a Q38.

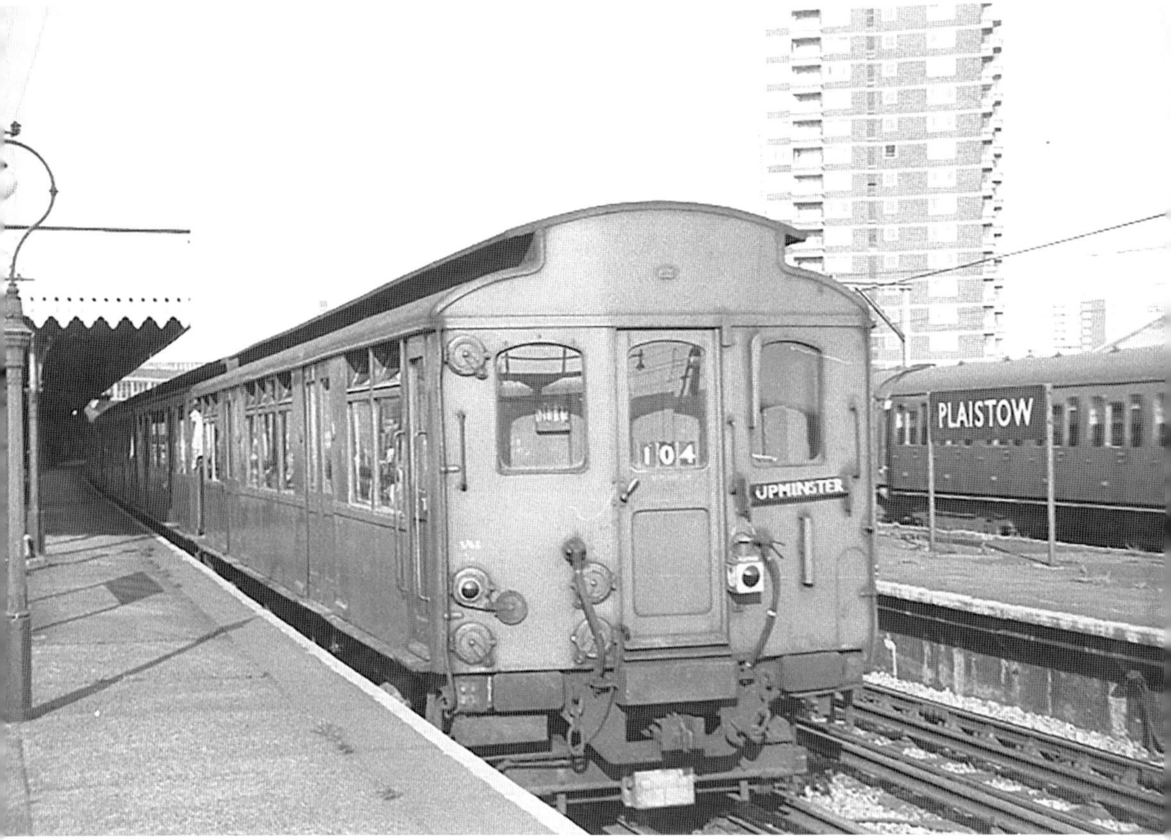
104
UPMINSTER
PLAISTOW

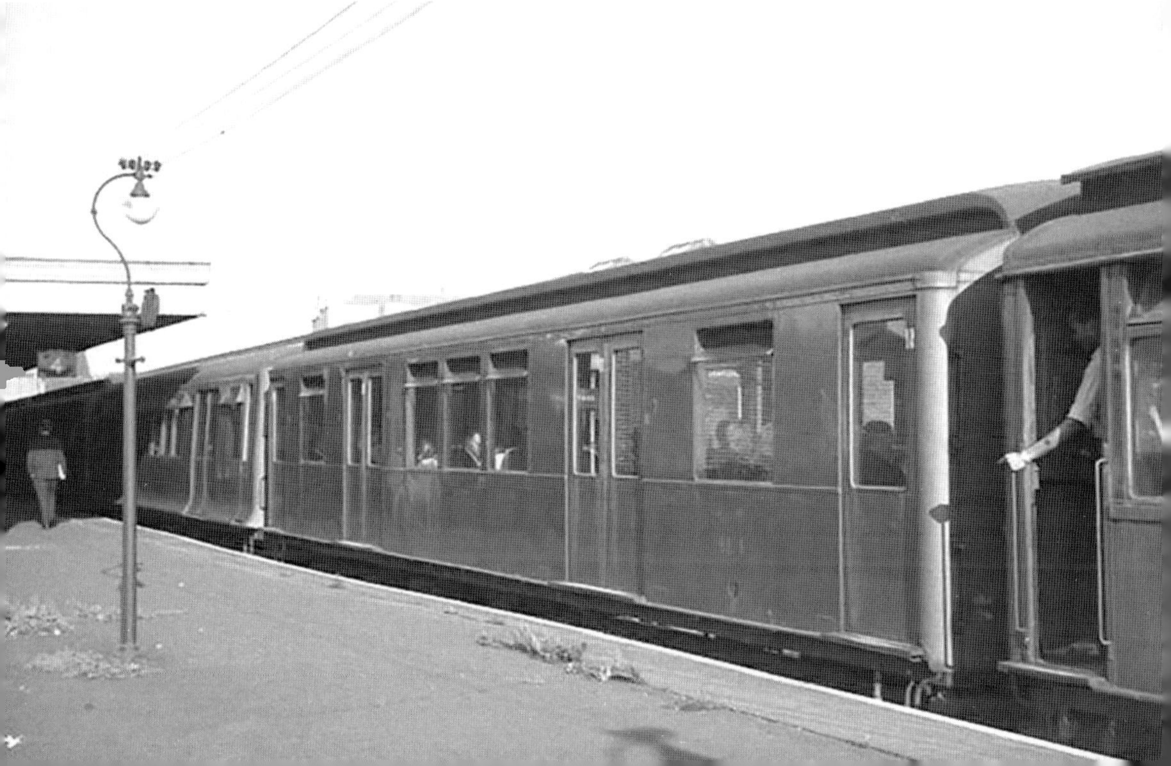

Above: **Evening shadows** lengthen as District Line set No.104 calls at Becontree on its way from Upminster to Ealing Broadway. It is the same train as illustrated at Plaistow earlier, with Q23 driving motor car 4248 leading. It appears at first sight to be entirely composed of clerestory-roofed Q Stock, however there is a Q38 lurking towards the end of the formation – such was the complexity of Q Stock. On the right are typical London County Council houses in their huge Becontree Estate, built in the inter-war years. Happily, Q23 car 4248 survives today, as a static exhibit in the LT Museum.

Opposite above: **The guard** of CO Stock driving motor car 54023 looks out from its cab, as it departs from Moorgate Station at the trailing end of an 'inner rail' (anti-clockwise) Circle Line train on 10 August 1969. The days of these units on such services are now numbered, with their impending replacement by new C69 Stock units. Most would see further use on the District and East London Lines, where within eighteen months or so they would see off the last remaining Q Stock.

Opposite below: **A visit** to Lillie Bridge Depot on 16 August 1969 finds former Great Western 0-6-0PTs L89 (left) and L94 (right) outside their shed, both being in steam at the time. Confusingly, the place where they had had their B.R. numbers on the smokebox door was used by LT for their duty numbers, i.e., 510 on loco L89. This dated from 1929 and had been GWR No.5775, acquired in 1963. L94 dated from 1930, having been No.7752, acquired in 1960.

Above: **On the** same occasion, Battery locomotive L39 is at the depot. Such locos, which replenish their batteries from the fourth rail, are used in battery mode when hauling engineers' trains around the system during non-traffic hours when the current is switched off, as well as on sections of non-electrified track. Having a standardised general design, this is one of the first batch of nine such purpose-built locos, supplied by the Gloucester Railway & Carriage Works in 1936/37.

Opposite above: **The same** general design was used for two new battery locomotives supplied by Metro-Cammell in 1964, L20 and L21. The former is dwarfed by the Earls Court exhibition hall when standing outside Lillie Bridge Depot alongside the West London Line. The latter was well-known at this period for hauling the Northern City Line stock transfers over the otherwise disused ex-GNR/LNER branch between Highgate and Finsbury Park over which the line should have been extended, as illustrated earlier. Another batch of eleven such locos followed in 1965, with others being delivered in the early 1970s. They remain in use at the time of writing.

Opposite below: **For many** years it was common practice on the Underground to convert withdrawn passenger stock for use on engineers' trains, for depot shunting and so on. In the 1950s and 1960s, many Standard Stock driving motor cars were converted for such use, some surviving into the 1990s. One such at Lillie Bridge is L69, which is a ballast motor car and attached to a couple of flat wagons with L20 at the other end of the formation.

THE LONDON UNDERGROUND, 1968–1985 • 45

Another Standard Stock ballast motor car at Lillie Bridge that day is L72. Whereas the usual livery for Underground non-passenger rolling stock at this time was maroon, this one is still in an earlier grey livery, having been converted in 1954.

Some Standard Stock trailers were retained too for use in engineers' trains as personnel carriers, needed when large scale track possessions, for example the closure of entire sections of line for track renewal, required large numbers of workmen. These cars were hauled either by Standard Stock driving motor cars, or battery locomotives. One of these, PC852, is inside Lillie Bridge Depot. It had previously been car 7080 in the passenger fleet and was demoted in 1966.

On Sunday, 24 August 1969, a special tour to commemorate the centenary of the opening of the District Railway was organised. Dubbed 'The District Centenarian', it was operated by a train of all clerestory-roofed Q Stock cars of traditional District Railway design. These were nicely polished up, and Q27 driving motor car 4357 is at the head of the train when awaiting departure from Wimbledon Station. By way of contrast, an R Stock train is in the adjacent platform.

Armed with one of London Transport's excellent value-for-money Twin Rover tickets, which allowed a day's unlimited travel on all their Central Area (red) bus routes and throughout the Underground, as well as times for the tour provided by a friend at 55 Broadway, I pursued it in order to catch both still and cine-film at various points it covered. It visited such places as the Metropolitan main line to Amersham, the Met's Uxbridge branch and the former District Railway branches now served by the Piccadilly Line between Rayners Lane and Ealing Common, and between Acton Town and Hounslow West. This view finds Q23 driving motor car 4204 heading the tour through Boston Manor Station on the latter branch. It is interesting to reflect that cars of this type served the District Line for nearly half of its first century.

Above: **I finished** up using my Twin Rover that day to visit the Drayton Park terminus of my local Northern City Line, where an entire seven-car formation of 1938 Stock, headed by driving motor car 11169, is surely over-provision for the handful of passengers using the line that Sunday. Interestingly, the train number, 271, is the same as that of the bus route that ran literally directly above the line between 1960 and 2023 from Highbury & Islington to Moorgate. The Northern City Line was truncated at Drayton Park in the autumn of 1964, and its Standard Stock, the last full units of such in LT passenger service, replaced by 1938 Stock two years later.

Opposite above: **The Northern** City Line's platforms at its deep-level Finsbury Park terminus were taken over by the Piccadilly and Victoria Lines in order to provide cross-platform interchange between them in both directions when the latter opened. However, when this was first proposed in the early post-war years, it was assumed that by the time it happened, the Northern City Line would be using the high-level platforms at Finsbury Park, on the way to its extension to Highgate, Alexandra Palace, East Finchley and High Barnet. But, of course, it never happened after all. This view at Drayton Park shows that the southbound connection between its intended surface route northwards was completed; it was brought into use in 1942 (at the height of the war), but only ever used for stock transfers during LT's tenure of the Northern City. The section switch boxes beside the track on the right still show the words 'southbound from LNER' in August 1969. The basic structure for the northbound ramp to the LNER was also built, though not fitted out with tracks. The Northern City had to wait until 1976, when British Rail's Eastern Region took it over, finally bringing the ramps into their intended use. Tracks on the northern one took a different route to that originally intended, burrowing beneath the East Coast Main Line and emerging on the west side of Finsbury Park Station.

Opposite below: **An unusual** quirk on Saturdays at this period was that some District Line trains ran around the north side of the Circle Line from Aldgate East to Edgware Road on Wimbledon services, rather than around its south side as usual. On 6 September 1969, heading set No.75, CP Stock driving motor car 53237 arrives at Moorgate Station. Directly above the train may be seen a bombed-out building from the blitz, still there nearly thirty years later.

Above: **On the** same day as the previous picture, Q27 driving motor car 4327 has arrived at New Cross Gate, despite showing 'New Cross' as its destination. Clearly visible in the picture at the bottom left-hand corner of its cab front is its 1964 overhaul date, and classification. It was standard practice to show cars' overhaul dates in this position. Generally, it was expected that all Underground cars would have a full overhaul about every seven years at this period, or if due for withdrawal, be withdrawn by the time seven years was up. All Q Stock, including the newer Q38 cars, was indeed withdrawn by the spring of 1971, the East London Line being their last refuge. Also of note on the right is one of British Rail's ubiquitous English Electric 350hp 0-6-0 diesel shunters, of which more than 1,000 were produced in the 1950s and early 1960s, sweeping steam locomotives from their path.

Opposite above: **Complete with** original tungsten lighting and decorative lampshades, this is an interior view of Q23 driving motor car 4204, one of London Transport's oldest passenger rolling stock in use at the time the Greater London Council assumed overall responsibility for its Underground network. Interior decor was of a pleasant green and cream, common to all tube and open saloon sub-surface cars until the appearance of the 1956 tube stock. It awaits departure from New Cross Gate to Whitechapel, also on 6 September 1969.

Opposite below: **One of** few oddities on the Underground system is Bakerloo Line Standard Stock trailer 70518, at Stanmore terminus on 8 September 1969. It is one of two of the '58 trailers' marshalled into 1938/49 Stock sets that are of the later type of this stock, with passenger doors at each end. This reduced their seating capacity by eight seats, compared to the older cars of the type which did not have the extra doors. This car survived in service until 1972, after which it was retained as a personnel carrier numbered PC856 and used with engineers' trains.

Above: **On 29 September** 1969, a six-car formation all-clerestory-roofed Q Stock has terminated in the bay platform of Putney Bridge Station in the evening rush hour. The trailing driving motor car for its next journey, to Upminster in the evening rush hour, is Q27 4289. These will end their District Line careers working during rush hours only, finally being confined to the East London Line from where the last was withdrawn in February 1971.

Opposite above: **Wet and** gloomy weather on 16 November 1969 provides a fitting context for this view, looking south, of the structural steelwork for the imposing new facade and additional two high-level platforms at Finsbury Park Station. This was meant for the extension of the Northern City Line over the LNER Northern Heights branch to Highgate and Alexandra Palace, along with rush hour journeys to East Finchley and High Barnet. To accommodate it, Station Place had to be moved eastwards, necessitating the demolition and rebuilding of a parade of shops and a pub facing the station. Opposite the station's northern end, a large furniture shop had to be demolished too, in order to provide space for the extra bridge span carrying the new Underground tracks across Seven Sisters Road. And all was thrust to waste when the extension was abandoned in the early 1950s. The station remained a rusting eyesore until the structure was demolished in 1972, but the 'temporary' wooden fencing erected in 1939 was not removed until the forecourt was finally tidied up some ten years later.

Opposite below: **Similarly, much** work was wasted on the extension of the Northern Line over the former LNER branch from Mill Hill East to Edgware. Captured on an equally dreary 23 November 1969 is the electricity substation for it at Page Street, alongside which cable runs for the electrification are still complete along both sides of the trackbed, which had been widened from single-track for the scheme, necessitating several bridge reconstructions. Officially, this part of the scheme was only 'in abeyance', until finally killed off by the M1 Motorway in 1964, as mentioned earlier.

THE LONDON UNDERGROUND, 1968–1985 • **53**

Above: **For the** benefit of members of the London Underground Railway Society on an official visit to Neasden Depot on 18 April 1970, ex-GWR/BR Western Region 57XX Class 0-6-0PT L90 has been nicely cleaned up and posed for our cameras. A contrast somewhat from its appearance when illustrated previously in these pages.

Opposite above: **The former** Metropolitan Railway Neasden Depot was considerably modernised as part of LT's 1935-1940 New Works Programme for the Bakerloo Line's 1939 takeover of the Stanmore branch. 1938 Stock cars with the aid of the Standard Stock '58 trailers' were allocated there for this, with extra new cars termed 1949 Stock added into their formations after the war. Also on 18 April 1970, '58 trailer' 70536, between 1938 driving motors 11117 and 10117, undergoes attention inside the modernised part of the depot.

Opposite below: **The oldest** working motorised rolling stock on the Underground at this period were electric sleet locomotives, used for de-icing the rails during winter. They were each constructed from two Central London Railway driving motor cars dating from 1903, joined back-to-back. ESL106 is one of two of these standing outside Neasden's modernised sheds. It is of note how the design of the sheds here is very similar to that of Hainault and West Ruislip depots built for the Central Line extensions under the same programme, not to mention Aldenham Depot built but never used for the Northern Line's extension to Bushey Heath. This, of course, was adapted to become LT's main bus overhaul works after the extension's abandonment, having been used during the war to build aircraft for bombing Nazi Germany into submission.

THE LONDON UNDERGROUND, 1968–1985 • **55**

Above: **Also present** in the sidings at the north end of Neasden Depot is electric sleet locomotive ESL107. This one survived in use well into the 1980s and is today preserved at the LT Museum's Acton Depot. Beside it on the right is 1938 Stock driving motor car 11134, wrecked in a collision in which its driver was unfortunately killed.

Left: **Another elderly** resident of Neasden Depot that day is Metropolitan Railway electric locomotive No.1 *John Lyon*. It is one of twenty built for hauling passenger trains on the Metropolitan's 'main line' between Liverpool Street, Baker Street and Rickmansworth. Essentially new locos supplied in 1922/23 by Metropolitan Vickers, they included some parts from earlier ones new in 1906/07 which they replaced. Withdrawal from regular passenger use came in the autumn of 1961, and this was one of three retained for such duties as depot shunting. Unfortunately, unlike the other two, it did not survive into preservation.

Several of London Transport's battery locomotives were present in Neasden's sidings too. This is L55, one of three supplied by Pickering in 1951. A further four were built by them in 1962. It is in an engineers' train formation of two Standard Stock trailers in use as personnel carriers, with another battery locomotive at the far end.

Of the same general appearance, battery locomotive No. L35 is the first of nine delivered by the Gloucester Railway Carriage & Wagon Company in 1936/37 and is coupled to some flat wagons. It too is preserved today at the LT Museum's Acton Depot. It contrasts with an A Stock driving motor car on the right.

Above: **L22 was** the first of a batch of eleven battery locomotives of the same basic design supplied by Metro-Cammell in 1965. Of particular note on these locos are their dual-coupling arrangements, with couplers to fit tube stock nearest track level, and hinged buffers folded inwards towards the cab in these pictures along with the relevant equipment to couple with main-line size rolling stock above them. This one is on the other end of the formation of personnel carriers, illustrated above.

Opposite above: **The personnel** carriers formed up between battery locomotives L22 and L55 are PC854 and PC855, later Standard Stock trailers built in the early 1930s, and adapted for such use after withdrawal from the Northern City Line in 1966. Other such cars were 'exported' to the Isle of Wight, for its surviving line between Ryde and Shanklin's electrification the following year, some surviving in service until as late as 1990.

Opposite below: **The battery** locomotive usually employed on stock transfers between the main Northern Line complex at Highgate Depot and the isolated Northern City Line in the final years of their traversing the uncompleted extension was L21, one of the two 1964-built Metro-Cammell examples. On 1 July 1970, it approaches the disused Crouch End Station. But instead of the red 1938 Stock usually involved, it is towing silver cars. Also of note in this picture is the never-used cable racking for the electrification on the left, the dilapidated footbridge linking footpaths either side of the railway and, on the right in the background, the sub-station at Crouch Hill.

***Above*: In contrast** to the Northern Line's extensions over LNER metals within the 1935-1940 New Works Programme, those of the Central Line (also halted during the war) were completed after it. Part of them was the inclusion of the Hainault Loop in the Central Line. However, only the section from Woodford via Hainault to Newbury Park was actually to be used, the remaining part from the latter point to Ilford and Seven Kings, where it originally joined the Great Eastern main line, was eventually abandoned. This view on 12 July 1970 shows a westbound eight-car train of 1962 Stock bound for Ealing Broadway descending into the Central Line's newly-built tunnel section at Newbury Park, from which it will emerge at Leytonstone. The two tunnels are positioned either side of the original GER/LNER line towards Ilford, by now reduced to single-track as this picture shows. It was used for some while after closure to passengers for goods trains, and the Central Line's unused tunnels had been put to good use during the war for making various materials for the war effort.

***Opposite above*: A London** Underground Railway Society visit to the District Line's Ealing Common Depot on 26 September 1970 finds Metropolitan electric locomotive No.12 *Sarah Siddons* undergoing refurbishment; one of its bogies has been removed and is in the foreground. This loco, of course, has subsequently been restored as a working exhibit attached to the London Transport Museum.

***Opposite below*: The days** of Q Stock in service are now very much numbered, but many of them are still in evidence at Ealing Common Depot on this occasion. In the sidings at its eastern end, nicknamed 'The Alps' by District Line staff, Q27 driving motor cars 4191 and 4235 are at the eastern ends of formations which appear still to be in use at the time. A CO/CP Stock driving motor car is visible on the extreme right.

THE LONDON UNDERGROUND, 1968–1985 • **63**

Above: **At the** western end of one of the formations of Q Stock in the previous picture is Q23 driving motor car 4200, one of the very oldest cars still in passenger service on the Underground at this time. Such use as Q Stock still had on the District Line was limited to rush hours only, although they still ran the entire East London Line service, though only for another five months. Perhaps some of the Q Stock here was retained for spares pending their final demise. Today, preserved Q Stock cars reside here in the part of the depot complex used for the LT Museum's road and rail exhibits, some of them in working order.

Opposite above: **An even** older resident at Ealing Common Depot that day is electric sleet locomotive ESL111, composed of the leading ends of two former Central London Railway driving motor cars dating from 1903. The R Stock car on the left shows just how Underground rolling stock design had changed in just 35 years or so.

Opposite below: **On the** same stretch of track as ESL111 inside Ealing Common Depot is former Standard Stock driving motor car L65, another of these converted for use as a ballast motor car after withdrawal. This view of its trailing end shows how some of its windows have been plated over, no longer being needed now that it is no longer in passenger use. It too contrasts with the R Stock motor car on the right, their initial body designs being only some fifteen or so years apart.

THE LONDON UNDERGROUND, 1968–1985 • **65**

Above: **Q23 Stock** driving motor car 4198 is one of two examples standing at the western end of Ealing Common Depot's car sheds. With peeling paintwork and faded fleet number and LT transfers, it has clearly seen better days. However, it had in fact been withdrawn from passenger service some years previously and is in use as a depot shunter.

Opposite above: **In much** better shape, fellow Q23 Stock driving motor car 4252 is surrounded by units of R Stock in the sidings between the car shed and the District and Piccadilly running lines. It has been in rush hour service the previous day, but sights such as this will soon be a memory at Ealing Common Depot.

Opposite below: **In the** sidings on the north side of the car sheds, CP Stock driving motor 54223 is at the trailing end of a three-car formation freshly overhauled at nearby Acton Works. Of note is the line name plate saying 'Metropolitan' on display, showing that it is probably destined for the East London Line where such units would see off the very last Q Stock. Such CO and CP Stock units were now rapidly being displaced by new trains of C69 Stock on the Circle and Hammersmith & City Lines.

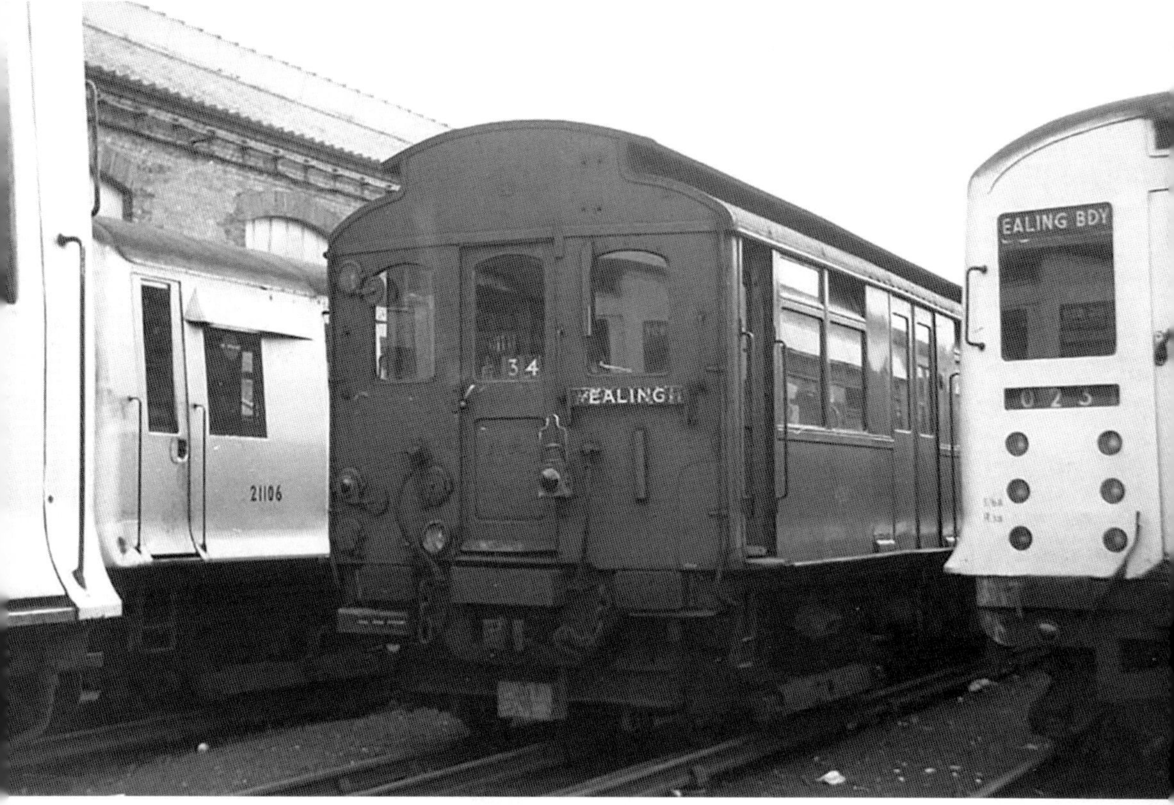
21106
34
EALING
EALING BDY

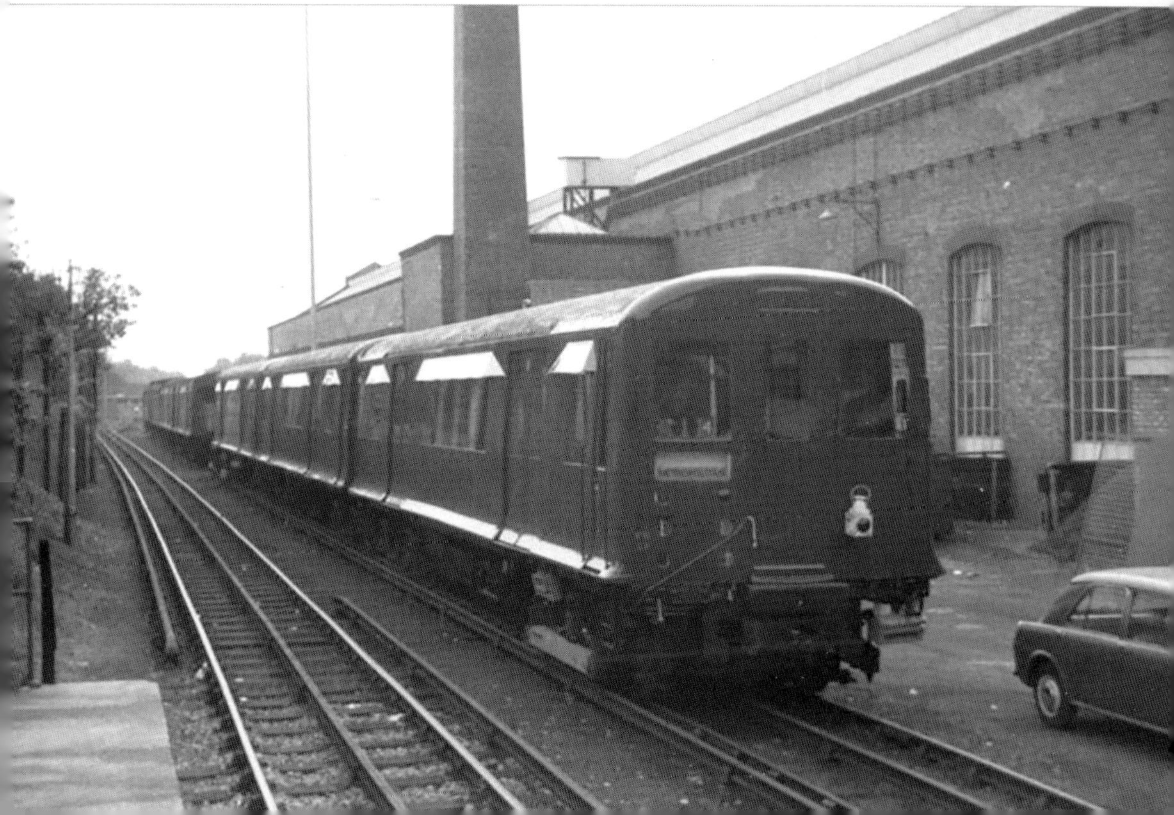

Above: **A real** oddity tucked away inside Ealing Common Depot is a two-car unit of 1935/36 Experimental Stock, comprising driving motor cars 11001 and 10001, which after withdrawal had been experimentally converted to an articulated unit. This was to assess the feasibility of building units of that configuration for future delivery. The idea was not perpetuated, and the unit saw occasional use as a depot shunter before being scrapped. For a few years, it had been used on the Central Line's Epping to Ongar shuttle, also sometimes putting in an appearance on the Woodford to Hainault shuttle whose destination it still shows. The basic design of these cars, of course, was perpetuated in the 1938, 1949, 1959 and 1962 Stocks and when, latterly, painted in silver livery, they could at first sight be taken for the latter, only the old-style destination code indicator lights on the left-hand side of the cab giving the game away.

Opposite above left: **One of** the stations the 1935/36 Experimental Prototype Stock served at the end of their passenger carrying careers on the Epping to Ongar shuttle was, of course, the little-used Blake Hall station. At this period, the entire branch was threatened with closure owing to the Greater London Council's desire to shed their newly-acquired London Transport services which strayed beyond their administrative boundary, especially when the local authority whose area it served, Essex County Council, was reluctant to subsidise it. So on 28 September 1970, I took the afternoon off from 'work' (by coincidence in the GLC's Planning & Transportation Department's offices at County Hall) to visit and photograph the line. This view of me standing by the sign directing (hopefully) passengers to the station that bears my family name, taken for me by my old friend Ken Wright, illustrates how remote the station was, and how incongruous the LT sign looks in the middle of nowhere in rural Essex. The main road behind me is that between Epping, North Weald and Ongar, which at the time had London Country bus route 339 (inherited from LT on 1 January that year) providing a decent service that completely duplicated the Central Line branch.

Opposite above right: **On that** afternoon, 1962 Stock driving motor car 1455 heads a four-car unit approaching Blake Hall Station on its way to Epping. Although the branch was always single-track, there would probably have been sufficient space to install a second track if intended pre-war plans for London 'overspill' housing development had been fulfilled, and perhaps a through Central Line service from Ongar to Central London operated.

THE LONDON UNDERGROUND, 1968–1985 • **69**

Below: **The train** in the previous picture (with driving motor car 1454 at its trailing end) calls at the sleepy Blake Hall Station, and a lady passenger on the platform appears to be just in time to catch it. Probably she knows the exact time this mid-afternoon train is due and is one of the station's regular handful of passengers. Meanwhile, the station manager's family's washing is hanging out to dry on this warm, sunny autumn day.

Above: **Still sporting** a splendid set of semaphore signals beyond its country (eastern) end, North Weald Station was provided with a second platform following takeover by London Transport in order for short-working trains to terminate there in rush hours. In this view, the train seen in the previous two pictures has now been to Epping and then back to Ongar, and, still with driving motor car 1455 leading, approaches North Weald Station on another trip to Epping.

Opposite above: **North Weald** Station boasted the very last level crossing on the Underground system at its country end, which however saw little use since it served a farm track. It is at the country end of the station, and also accommodates a public footpath. With the cab door still open, driving motor car 1454 passes over it at the train's trailing end as it arrives at the station. One of LT's standard notices warning of the dangers of trespassing is attached to the crossing gate.

Opposite below: **Rush hour** at North Weald Station. When one of the short-working units terminated at North Weald Station, a picture like this could be taken of one on the through service alongside it. Here, the unit illustrated in the previous pictures is in the bay platform, whilst another four-car 1962 Stock unit, with driving motor car No.1462 nearest the camera, is on the 'main' line. The threat of closure would have meant the end of such scenes, but in the event, Underground trains continued to work out to Ongar until the end of September 1994.

Above: **One of** a number of long-closed stations on the London Underground is City Road, one of the early City & South London Railway stations. It was closed in 1926 in conjunction with structural alterations to the C&SLR stations and tunnels to allow standard size tube trains to work through them upon the line's incorporation into what was later to become the Northern Line complex. On a murky New Year's Day 1971, part of the station's surface buildings at the junction of Central Street and City Road itself have recently been demolished in conjunction with the erection of the monstrous Islington Council tower block on the right. Part of the building survives today, however, since it accommodates a ventilator shaft to the tracks below.

Opposite above: **On 6 June** 1971, a special run was arranged to mark the final withdrawal of steam locomotives from the London Underground system. A special train hauled by one of the former BR Western Region 5700 Class pannier tanks, composed of typical wagons and brake vans used in their final years of service, ran from Moorgate Met. to Neasden Depot. Some of this route, of course, the section between Farringdon and Baker Street, had been part of the very first Underground line, the 'cut and cover' Metropolitan Railway back in 1863. In the morning of that final day, loco L94 emerges from the Met. main line into Baker Street Station on the way to collect its train in one of Moorgate's terminal platforms.

Opposite below: **An open** day was held at Neasden Depot in conjunction with the event, and special trains ran taking spectators to it. One such is formed of recently-delivered C69 Stock, which was then replacing CO/CP Stock on the Circle and Hammersmith & City Lines. Driving motor car 5561 speeds past West Hampstead Bakerloo Line Station on the Met. main line as people waiting there for the steam special to arrive look on. This new train itself is, of course, history now and I am glad I did not ignore it. The last of these units was withdrawn in 2014.

Above: **The steam** special is not far behind, and to the annoyance of those of us wanting a clear photograph of it passing West Hampstead Station, it is on the Bakerloo local line and not the Met. main line. Its headboard leaves no doubt as to what it is commemorating. To a great extent, this event was more important to Britain's railway history than the withdrawal of British Railways' last steam locomotives in the summer of 1968; after all, London Transport's pannier tanks were also standard gauge locos employed by a publicly-owned, passenger-carrying railway. That is sadly often forgotten today.

Opposite above: **At Neasden** Depot, alongside the Metropolitan and Great Central main lines, fellow pannier tank L90 is at the London end of L94's train, as crowds of visitors view them. Both of these final survivors of London Transport steam power were saved for preservation.

Opposite below: **As is** standard procedure at depot open days, various interesting items of rolling stock were on display at Neasden that day. Nearest the camera here, electric sleet locomotive ESL100, composed of two 1903 vintage ex-Central London Railway driving motor cars joined back-to-back, has been smartened up for the occasion. Behind it, Metropolitan Railway electric locomotive No. 12 *Sarah Siddons* carries a headboard explaining how it is in use for brake block testing.

Above: **London Underground's** battery locomotives are not often seen by the general public, being used as they are usually in the dead of night during non-traffic hours. However, on this occasion, one of them, L18, is proudly displayed with the elderly electric locomotives in the previous two pictures. Although its general design dates back to the mid-1930s, it is in fact brand new, one of three supplied by Metro-Cammell in 1971.

Opposite above: **A very** quaint exhibit at Neasden is the two-car electric sleet 'locomotive' ESL118 & 118A, formed of two former T Stock driving motor cars, 2758 and 2749 respectively, which powered compartment-stock Metropolitan Railway electric multiple units. Such units, newly-built for electric operation, dated from 1927 to 1932, though some had been rebuilt from former steam-hauled stock from 1920 onwards. The last ran in service in October 1962. Used initially for de-icing, the two cars were later used in experimental trials as a leaf removal unit later in the 1970s. Fortunately, both were preserved after final withdrawal.

Opposite below: **Returning to** current Underground passenger rolling stock, on 2 July 1971 a four-car unit composed of two 1960 Stock driving motor cars and two Standard Stock trailers has just departed from Chigwell Station into the deep cutting to its east on the Hainault Loop. Car 3909 is at its trailing end. At this period, these units, with the exception of the one used for track-testing purposes, were employed on the Woodford to Hainault shuttle service on this quiet section of the Central Line.

Above: **Owing to** its quiet nature, this section of the Central Line was used as a test-bed for automatic train operation on the Underground, with new 1967 Stock units for the Victoria Line being tried out on it prior to entering service in that mode on the new line in 1968/69. Some of these remained in use there until the Victoria Line was extended from Victoria to Brixton (on 24 July 1971), there being a few units to spare. The next train along heading east from Chigwell is one, with driving motor car 3017 at its trailing end.

Opposite above: **Another Victoria** Line 1967 Stock unit, with driving motor car 3024 trailing, has called at Chigwell Station on the same occasion. Built by Metro-Cammell, these units never saw public service in daylight other than when used on the Hainault Loop. It is of note that the station is still in basically original Great Eastern Railway condition, as are all of those on the Hainault loop (except Hainault itself, as illustrated earlier) as well as those on the 'main' Central Line between Leyton and Epping, with the notable exception of Loughton.

Opposite below: **On 26 September** 1971, CP Stock driving motor car 53251 heads an 'outer rail' Circle Line train into Farringdon Station. By now, such units are rapidly being replaced by new C69 Stock and displaced to the District Line. Of note in this picture is the signal box behind the leading car on the left, and on the right beneath its trailing car, the junction with the Metropolitan Widened Lines may just be discerned. Although these lines were worked by steam-hauled and diesel trains, coming from the East Coast and Midland main lines, the tracks were electrified for Underground use for a few years in the 1920s and 1930s. At the time of writing, work is in progress to electrify them on LU's fourth-rail system along the disused section between Farringdon and Barbican in order to provide stabling facilities for S Stock.

Above: **On the** same day, CO Stock driving motor car 53015 heads a similar formation when departing from Paddington Station for Hammersmith on the Hammersmith & City section of the Metropolitan Line. The days of such workings here are numbered too, with their ongoing replacement by new C69 Stock. A curious feature illustrated here is that this line's tracks emerge from their tunnel into the former Great Western main line terminus, between its suburban platforms. They are a relic of when through workings, both passenger and goods from the GWR, ran on to the Metropolitan Railway's system, the latter continuing into the 1960s.

Opposite above: **A sad** sight on 18 March 1972 is the trackbed of the former GNR/LNER Northern Heights line from Finsbury Park to Alexandra Palace, looking towards Highgate east tunnels. Last used for stock transfers to and from the Northern City Line in the autumn of 1970, its tracks have recently been lifted. Much evidence of wasted works on the abandoned Northern Line extensions remains, not least the completely rebuilt Highgate High Level Station at the far end of the tunnel, as well as infrastructure such as cable runs for the electrification, still clearly visible here on both sides of the formation.

Opposite below: **Thanks to** a friend who was working there as an electrician, I paid another visit to Ealing Common Depot on 15 June 1972. As usual, an interesting selection of rolling stock was present, not least Metropolitan Railway electric locomotive No.5 *John Hampden*. At the time, it is being used for shunting purposes and on engineers' trains. Happily, it was later preserved and is part of the LT Museum collection today.

In contrast, a sad sight within the depot car shed is former Standard Stock driving motor car L69, one of those converted for engineers' and shunting use. As its buckled underframe shows, it has been damaged in a collision and is now being broken up, with various parts being salvaged to keep others of this type in use.

The other victim of the collision was similarly-converted former Standard Stock driving motor car L62, damage to which is not quite so severe. It awaits its fate at the rear of the depot, on 'the Alps' sidings.

Standing forlornly next to the damaged L62, COP Stock trailer 013070 is also withdrawn. Following the final withdrawal of Q Stock and the transfer of all remaining CO and CP Stock from the Circle and Hammersmith & City Lines to the District Line, train formations were revised, meaning that some such cars were surplus to requirements.

Stabled amid a mixture of rolling stock in the sidings at Ealing Common that day is CP Stock driving motor car 53264, heading a unit which appears to be out of use. It is interesting to compare it to the R Stock driving motor car on the adjacent track; nearly all such R Stock cars were converted from Q38 Stock trailers, which (as remarked upon above) were designed with a view to such conversion at a later date. Unlike CO/CP Stock formations, which had trailers and driving motor cars, all cars of R Stock were motorised, those of the 'newbuild' R47, R49 and R59 varieties being non-driving motor cars except for a few R49s.

Illustrating how the District Line provided the rolling stock for the East London Line, despite its being part of the Metropolitan Line complex for many years, CP Stock driving motor car 54252 is amongst a group of these units stabled at Ealing Common Depot. Its line-describing plate indicates it has been in use there; curiously, such plates reading 'East London Line' did not exist at this time. These units had, of course, operated on the Metropolitan main line until replacement by new A Stock in the early 1960s.

Something of a surprise at this period was the withdrawal and scrapping of a number of relatively new R47 Stock non-driving motor cars. Car 23226 is nearest the camera in this sad line-up alongside Ealing Common Depot's care shed. The cars were surplus to requirements as a result of changes in the formation of trains of R Stock. They had an operational life of only about half the length of time, around twenty-four years, that the some of the last venerable Q23s achieved.

One of the oldest residents at Ealing Common Depot on 15 June 1972 is ESL113, another of the electric sleet locomotives formed from two Central London Railway driving motor cars of 1903 vintage, joined back-to-back with their passenger sections removed. In the foreground are some of London Transport's flat wagons used for transporting such things as rail lengths in engineers' trains.

Another elderly resident stabled at the depot's western end is Metropolitan Railway electric locomotive No.12 *Sarah Siddons* in use at the time as a brake block test loco. It provides an interesting contrast with the CO/CP Stock driving motor car beside it.

Above: **Peeping out** of the car sheds, battery locomotive L29 leaves no one in any doubt of its stock number. It is one of the batch of eleven supplied by Metro-Cammell in 1965.

Opposite above: **By the** spring of 1973, work was progressing on the new Fleet Line, which was to take over the Bakerloo Line's branch between Stanmore and Baker Street, and then proceed in new construction beneath the West End to Charing Cross as a 'first stage' and then, hopefully, be extended beneath the Strand and the City and then further to the east. Crucial in these plans was connection between its 'interim' terminus at Charing Cross with the nearby Bakerloo Line Trafalgar Square Station, and the Northern Line Strand Station. Both were to be renamed Charing Cross, and on 15 April 1973, one of the original entrances to the latter, at the southern end of the forecourt of the Southern Region main line station, is about to be closed. Behind it, the working site for the new combined booking hall and entrances to the station is visible. It would be just over six years later that the new line opened, renamed the Jubilee Line, and of course its eastward projection would eventually be very different from that originally foreseen, with this terminus effectively abandoned. The Bakerloo, Circle, District and Northern Line station hitherto called Charing Cross was renamed Embankment as part of the scheme.

Opposite below: **Over the** years, a number of interesting tours on the Underground system were organised for enthusiasts, often taking rolling stock to parts of the system it never normally served. A good example was the Metro Tube Tour on Sunday, 13 May 1973, on which a train of Bakerloo Line 1938 Stock ran over much of the Circle, District, East London and Metropolitan Lines. This view sees it heading west through Farringdon Station, with driving motor car 10093 in the lead. Of note are the disused warehouses adjacent to the Widened Lines platforms on the right.

Above: **Part of** the formation of the Metro Tube tour train was '58 trailer' 70525, one of the fifty-eight Standard Stock trailers converted in 1938/39 to run with new 1938 Stock for when the Northern and Bakerloo lines' older stock was replaced by these, and the latter also took over the Metropolitan Line service on the Stanmore branch. It too is pictured passing through Farringdon Station, and by now their days were numbered. The last was withdrawn a few months later.

Opposite above: **Having traversed** the Circle Line in an anti-clockwise direction, on the 'inner rail', as far as Tower Hill, the tour continued along the District Line, then via St. Mary's curve east of Aldgate East to reach the East London Line, both of whose southern termini it visited. In this view, driving motor car 11054 is at its head awaiting departure from New Cross Gate. Its appearance here was a foretaste of what was to come the following year, as will be explained later.

Opposite below: **The East** London Line's incumbent rolling stock at this time consisted of five-car formations of CO/CP Stock, one of which is 'topped and tailed' by driving motor cars 53029 and 54029 departing from Surrey Docks Station heading south. These units had replaced the very last Q Stock in February 1971, having been ousted by new C69 Stock from the Circle and Hammersmith & City Lines.

Above: **Meanwhile, the** Metro Tube tour has travelled up to Shoreditch and back on the East London Line, returned to the District Line via St. Mary's curve again and now heads eastbound through Whitechapel Station. Driving motor car 11054 is at its trailing end. In the northernmost platform, a train of C69 Stock, with driving motor car 5562 leading, has terminated on the Hammersmith & City Line. Its smart black roof and unblemished unpainted aluminium bodywork is in stark contrast to how these trains were besmirched by the curse of graffiti in the late 1980s.

Opposite above: **The tour's** very comprehensive itinerary took it much further to the east, out to the District Line's Upminster terminus, if I recall correctly. This view finds it passing through Upton Park Station, with driving motor car 10093 at its trailing end. After this, it continued out to the far north-western reaches of the Metropolitan main line.

Opposite below: **On 22 June** 1973, this lucky shot in Roding Road, Loughton captures two classic London Transport designs, both bodied by Metro-Cammell of Birmingham. A train of Central Line 1962 Stock (in which, however, some of the non-driving motor cars were built by British Railways at Derby Works) passes above 1952-built A.E.C. Regal IV 39-seater bus RF498. The low height of this, and another bridge on the same line, restricted local bus route 254 to single-deck operation.

THE LONDON UNDERGROUND, 1968–1985 • **91**

Above：Clearly showing Debden Station's original 1865 Great Eastern Railway eastbound platform, this view shows an Epping-bound eight-car formation of Central Line 1962 Stock, with driving motor car 1641 at the trailing end, on 29 August 1973. Essentially a follow-up order from the 1959 Stock built for the Piccadilly Line, these trains included some B.R. Derby-built non-driving motor cars, as mentioned above. This was because, initially, seven-car 1959 Stock formations were diverted for use on the Central Line in 1960-1962 in order to expedite replacement of its Standard Stock once it had been decided not to proceed with mass production of 1960 Stock. Because the Central Line needed eight-car trains, the 'extra' non-driving motor cars were built ahead of the rest of the 1962 Stock by British Railways and inserted into the 1959 Stock formations. When this stock was transferred to its intended home on the Piccadilly Line, the BR-built cars were put into their correct 1962 Stock units.

Opposite above: On 6 July 1974, a Northern Line train of 1938 Stock heads south from Edgware Station bound for Kennington, via Charing Cross. I took this picture from the bridge which once carried the GNR branch into its Edgware terminus. On the left is a train of 1972 Mk1 stock, whilst in the distance is a 1972 Mk2 unit. These were effectively two-person operated versions of the Victoria Line 1967 Stock, hastily supplied to the Northern Line as a result of it suffering from unreliability of its ageing 1938 Stock trains, being unkindly dubbed 'The Misery Line' by the media. Perhaps if the scheme to incorporate the Edgware to Mill Hill East and Highgate to Finsbury Park/Drayton Park portions of the uncompleted Northern Line extensions had been revived, making possible an alternative route from Edgware, Finchley and Highgate to the City, the line's overcrowding would have been alleviated. Sadly, my many letters to the local press advocating this at the time had no effect.

Below: **Almost five** years after I took the previous picture of the trackbed of the abandoned Mill Hill East to Edgware line at Page Street, the site has now been levelled and incorporated into the works complex of civil engineering contactor John Laing & Company Ltd. The never-used substation built for the line's electrification is just right of centre, the date is 11 July 1974.

Above: **This closer** view of the abandoned substation, which I took on 19 July 1974, shows its transformer chambers. The building was used for a while by Laing's for storage but was demolished not long afterwards. Identical buildings were erected at Crouch Hill (as illustrated earlier) and at Muswell Hill Station. Both were fully equipped too, despite comments to the contrary being published elsewhere, and it is probable this one also was; what a waste.

Opposite above: **Something of** a surprise at this period was the replacement of the East London Line's CO/CP Stock trains by 1938 Stock tube units. On 16 September 1974, driving motor car 10098 awaits departure from New Cross Gate Station on one of the rush hour journeys to Shoreditch. It is of note how the platform level matches the level of the train's floor. On the left in the sidings is one of British Rail's numerous English Electric 350hp 0-6-0 diesel shunters, freshly repainted and given its new '08' stock number.

Opposite below: **Two Central** Line 1962 Stock units meet at Epping Station on 20 September 1974, as driving motor 1684 arrives heading a four-car set on the shuttle service from Ongar. The other train has terminated on the 'main' line, having come through Central London from either West Ruislip or Ealing Broadway. Of note is the water tank beyond Platform One, a relic of when the shuttle was steam-hauled; such workings ceased in November 1957.

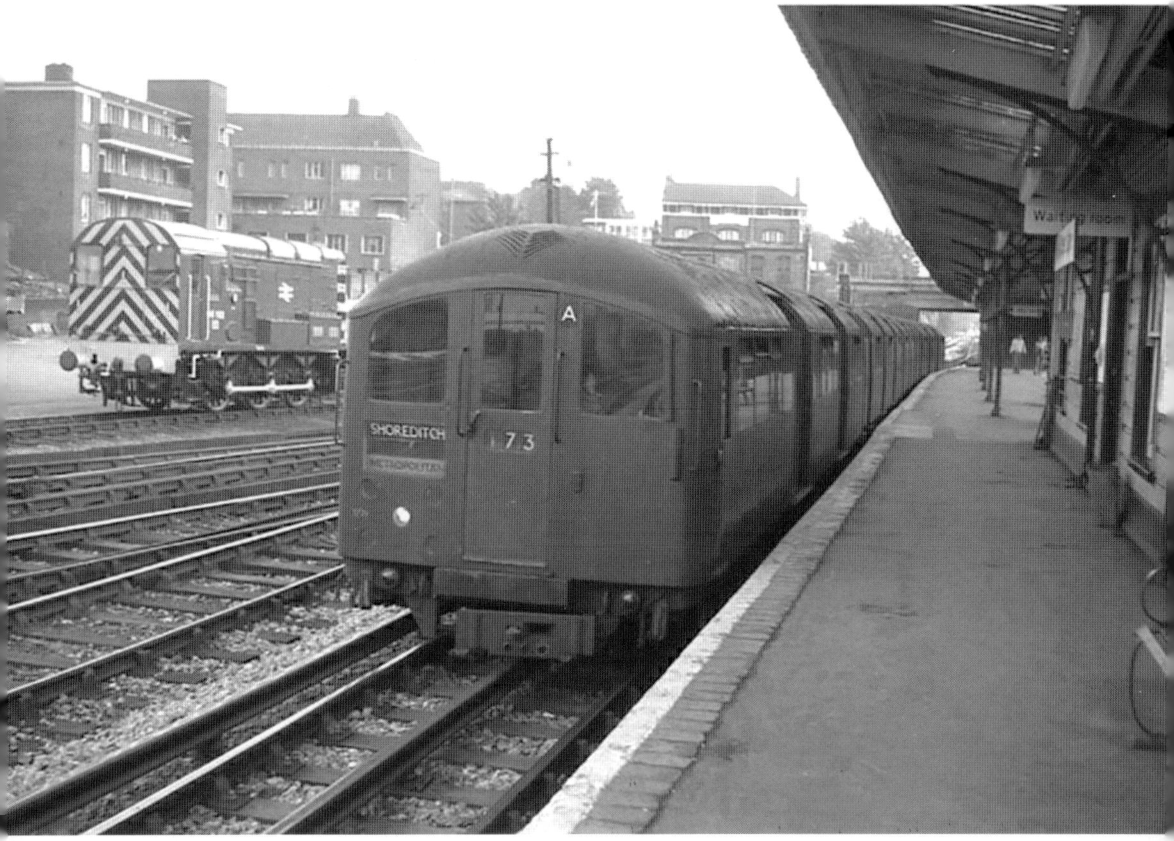
SHOREDITCH
73
A

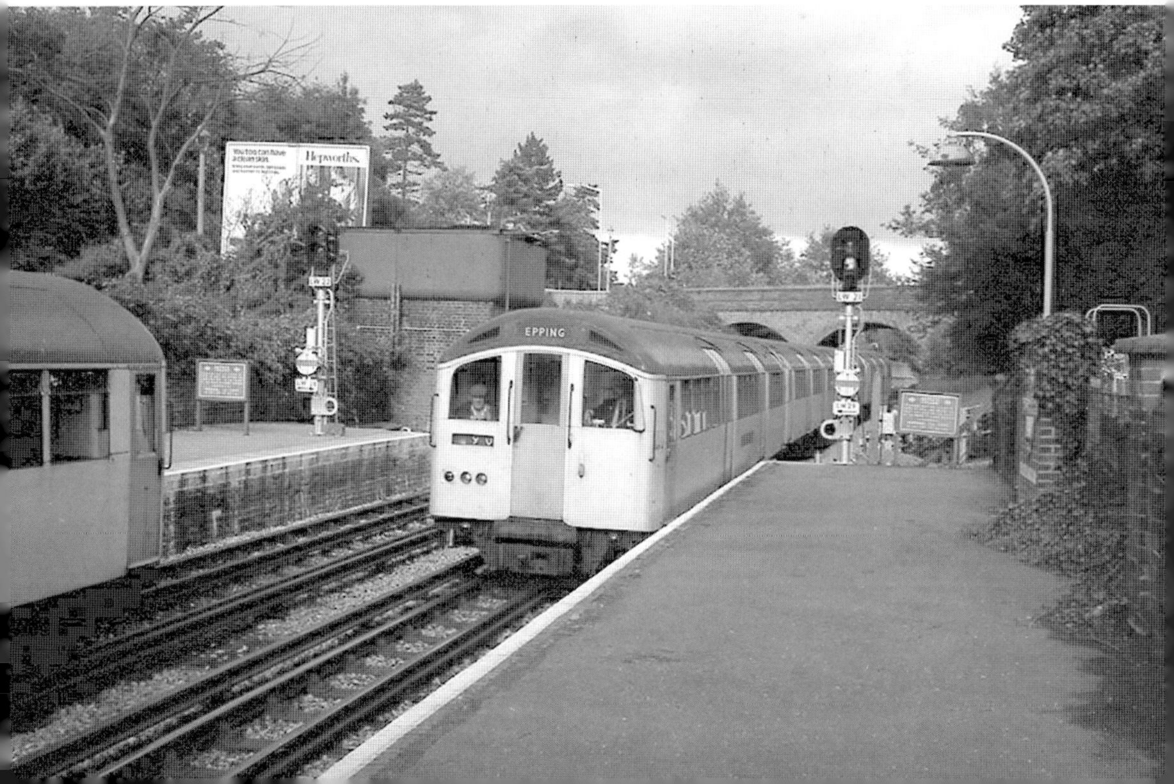
EPPING
Hepworths

Above: **A ride** on the shuttle unit took me to its Ongar terminus, where driving motor 1685 heads the train. This view is looking south from the main road beyond the station and illustrates how the line was originally constructed with a view to further extension, perhaps to rejoin the Great Eastern main line at Chelmsford.

Opposite above: **Back on** the Northern Line, driving motor car 10267 is at the trailing end of an Edgware-bound train departing from Hendon Central Station and entering the short tunnel north of it on New Year's Eve 1974. This car is one of those that were originally owned by the LNER, intended for use on the extension of the Northern Line over its tracks from Finsbury Park to Highgate, Alexandra Palace, East Finchley, High Barnet and Edgware via Mill Hill.

Opposite below: **On a** crisp, sunny winter's day, 3 January 1975, a train of Northern Line 1972 Mk2 Stock, headed by driving motor car 3561, has terminated from the north in one of East Finchley Station's centre platforms. These are on the alignment of the former GNR/LNER branch from Finsbury Park to Edgware and High Barnet and, following the abandonment of electrification of the section of line south of Highgate, have only been used for trains terminating here and running to/from Highgate Depot and Highgate Wood sidings. The splendid Charles Holden-designed station was built in 1939/40, replacing the original GNR structure, with trains on the tube extension of the Northern Line serving the other platforms and those (still steam-hauled) from Finsbury Park and the City using those in the centre. Of note too is the famous statue of the archer pointing his arrow towards the City, on the right of the picture. 1972 Mk2 Stock was distinguishable from the earlier Mk1s by having red passenger doors, and whereas all of those had been withdrawn by the end of 1999, most Mk2 units survive today on the Bakerloo Line, and may last as long as until 2040. They are already the oldest passenger railway rolling stock in regular, non-heritage use in the whole country.

Above: **Early on** the morning of 19 February 1975, the first complete six-car train of new 1973 Stock slowly emerges from Acton Works, passing Acton Town Station on test run. Headed by driving motor car 100, it displays a revised livery with the lower part of the cab front painted red, however the red-painted passenger doors have not been perpetuated. These units replaced the Piccadilly Line's 1938, 1956 prototype and 1959 Stock over the next couple of years and remain in use on that line today. Replacement is due within the next few years.

Opposite above: **One of** the worst disasters to befall London's Underground system took place at the height of the morning rush hour on 28 February 1975. A crowded train of 1938 Stock on the Northern City Line failed to stop at its Moorgate terminus, slamming into the short blind tunnel at the end of Platform Nine. Owing to the tunnel having been built to main-line dimensions, the first and second cars, and part of the third car, overrode one another and were jammed into the tunnel. Forty-three people perished as a result, with many more injured. At lunchtime on that fateful day, 1275LD, a London Transport railway emergency tender built on a Leyland Titan PD3 bus chassis, along with vehicles from the London Fire Brigade, are present outside the station's Moorgate Street entrance. Despite the fearful consequences of the crash, the Circle and Metropolitan and main Northern Line parts of the station were still open to passengers – something that would *never* happen nowadays in such circumstances.

Opposite below: **Such was** the devastation of the leading cars of the crash train at Moorgate that services there on the Northern City Line did not resume until several days later. Once the wreckage had been removed, a very long sand drag was provided at the end of platform nine, as illustrated in this view, with a 1938 Stock unit arrived there, a couple of days after reopening, on 8 March 1975. Of note here is the blackboard at the end of the platform, on which a plea for witnesses who may have been there when the crash occurred to come forward in an attempt to clarify what happened. Following the disaster, elaborate terminal protection measures were evolved to prevent such a tragedy happening again.

THE LONDON UNDERGROUND, 1968–1985 • **99**

Above: **In common** with those based at Drayton Park for the Northern City Line, 1938 Stock units working the East London Line (which had a small stabling depot at New Cross) had major maintenance carried out at Neasden Depot, which still housed such trains for the Bakerloo Line at the time. On 9 March 1975, driving motor car 11098 is at the trailing end of a four-car set heading through Baker Street Station along the Metropolitan Line on its way to Neasden on a stock transfer move.

Opposite above: **A visit** to the East London Line itself on 28 March 1975 finds a freshly-overhauled four-car set of 1938 Stock which has terminated at Whitechapel. Of note is the new London Transport livery recently adopted for both red-liveried Underground rolling stock and the bus fleet (with the exception of elderly RT and RF class vehicles). This saw the traditional gold stock numbers and fleet names replaced by white numerals and large white LT roundels. The unit is headed by driving motor car 10084, followed by trailer 012217, non-driving motor car 12090 and driving motor 11084.

Opposite below: **It was** indeed a pleasure to ride on this newly overhauled unit, whose interior had been repainted, its woodwork re-varnished and its seats given new moquette and leatherwork. With car 11084 leading, it catches the sun at New Cross terminus, confusingly showing the plate for the East London Line's other southern terminus, New Cross Gate.

THE LONDON UNDERGROUND, 1968–1985 • **101**

***Above*: In contrast,** 1938 Stock driving motor car 10190 looks somewhat careworn when awaiting departure with a seven-car Northern Line train bound for Kennington via Charing Cross at High Barnet terminus on 31 March 1975. A train of 1972 Mk1 Stock may be glimpsed in the sidings on the right. Proposals to partially rebuild this terminus as part of the 1935-1940 New Works Programme never came to fruition, and the station remains basically in its original GNR condition to this day.

***Opposite above*: In the** sidings adjacent to the car sheds at the Northern Line's Edgware terminus, 1938 Stock driving motor car 11277 is flanked by two units of 1972 Stock, also on 31 March 1975. It is another of the many cars of this type that were originally owned by the LNER, over whose branches from Finsbury Park to Highgate, Alexandra Palace, East Finchley, High Barnet and Edgware via Mill Hill the Northern Line should have been extended in 1939/40. Such cars still carried plates on their solebars reading 'Property of LNER' not only until the end of their London Transport careers, but also after their sale to British Rail for use on the Isle of Wight until only a few years ago.

***Opposite below*: Although my** attentions that day were devoted to trains of the Northern Line's 1938 Stock, now imminently due for replacement by 1956/59 Stock displaced from the Piccadilly Line, fortunately I also caught this view of a 1972 Mk1 Stock train departing for Edgware from Brent Station with driving motor car 3528 at its trailing end. Such units had much shorter working lives than the 1938 Stock, all being withdrawn by the end of 1999, although some cars from them were redeployed to work with 1967 and 1972 Mk2 Stock units – the latter still being in service today.

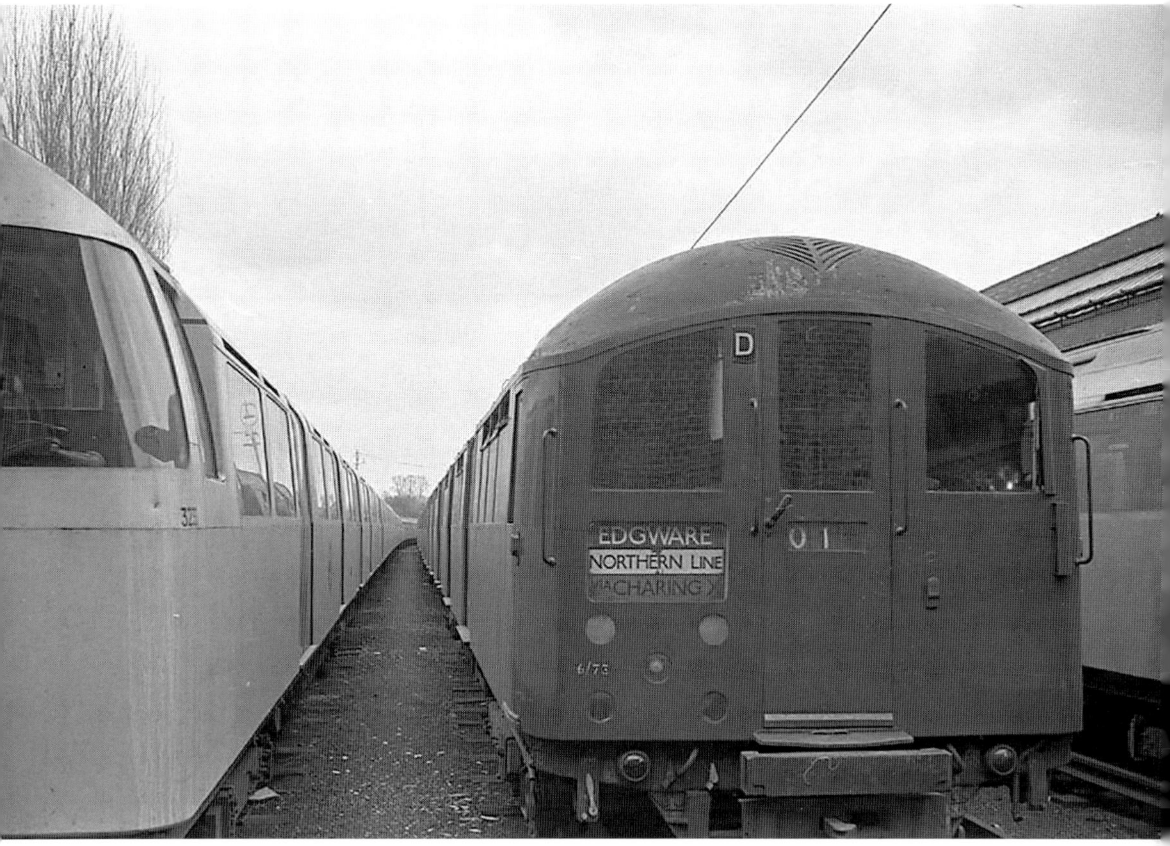
D
EDGWARE
NORTHERN LINE
CHARING X
01

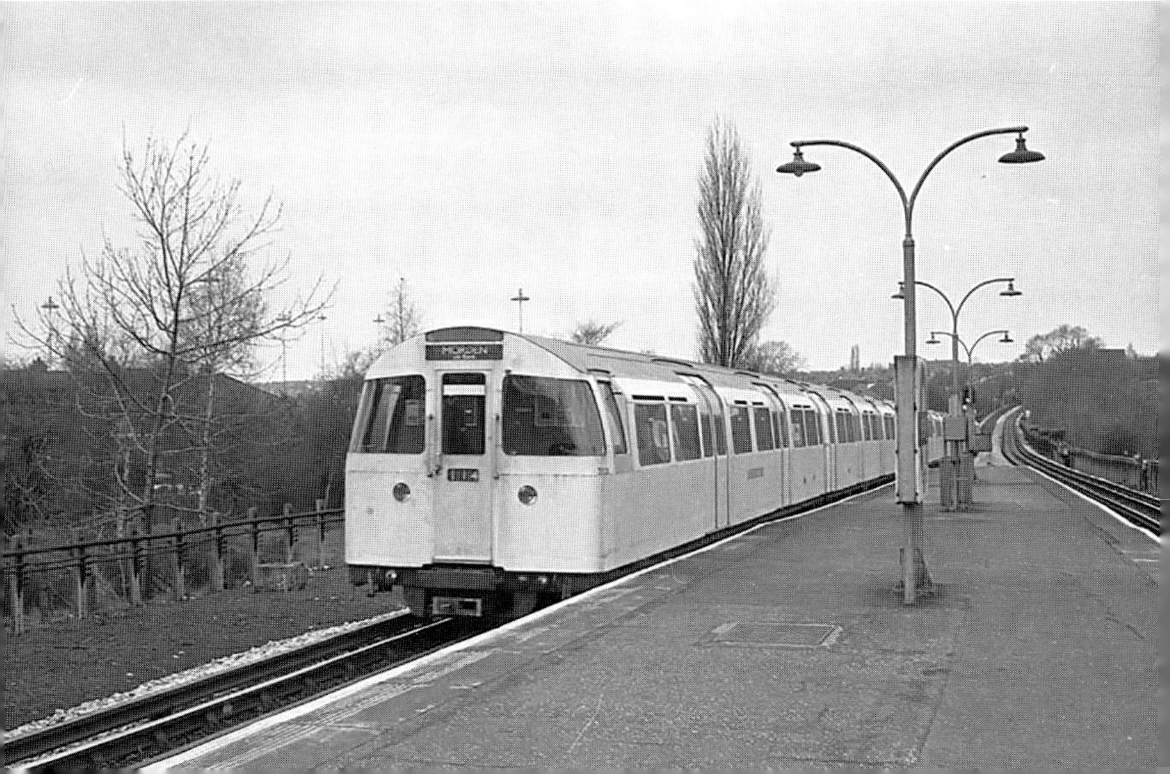

Above: **More typical** of Northern Line trains serving Brent Station at this time were, of course, 1938 Stock units. Driving motor car 10218 heads one of them arriving there with a service from Edgware to Morden via Bank. Of note here are the spaces either side of the track formation, a relic of the branch's early years when 'fast' services on the Edgware branch were operated non-stopping at this station, using two additional outer tracks. The station itself was renamed Brent Cross a few years later, following the opening of the nearby Brent Cross shopping centre in 1976.

Opposite above: **The uniquely** rebuilt 1938 Stock driving motor car 10306 is in the same formation at Brent Station, and this view amply illustrates its side windows extending into the roof. Small round windows, however, were provided at positions where the passenger doors opened inside the car. 1949 Stock trailer 12498 is behind it on the left. Eleven of these as well as seventy uncoupling non-driving motor cars, of the same general design as the 1938 Stock, arrived between 1949 and 1952 to increase the number of four-car units on the Northern Line (from three cars), trains thus working in seven-car formations made up of one three-car and one four-car unit.

Opposite below: **On 4 April** 1975, another visit to the Northern Line's northern reaches took me to Finchley Central Station, where 1938 Stock driving motor car 11221 leads a seven-car formation heading for Kennington via Charing Cross – the glass in the destination plate holder is apparently dirty, thus obscuring the latter information. This station, named Finchley (Church End) in steam days, was to have been rebuilt when the Northern Line was extended over the LNER Edgware and High Barnet branches (which diverged north of the station), with a new booking office on Ballards Lane above the station's country end. But as with so much of this part of London Transport's 1935-1940 New Works Programme, it never happened after all and the original GNR station buildings survive to this day.

THE LONDON UNDERGROUND, 1968–1985 • **105**

Above: **A similar** 1938 Stock unit, with 'LNER batch' driving motor car 10282 at its trailing end, carried me from Finchley Central to Woodside Park Station, which once again still has its original GNR buildings. These are of very similar design to those on that railway's branch from Wood Green (now named Alexandra Palace) to Enfield Town, which ironically had most of their original platform buildings drastically altered when the line was electrified in the mid-1970s, apart from Enfield Town of course which had been re-sited to become a through station when the line was extended to Hertford North.

Opposite above: **This view** on the same occasion shows the original booking office and stationmaster's house at Woodside Park, little changed from GNR days apart from having London Transport Underground signage. However, a level crossing taking Woodside Park Road across the country end of the station was removed upon electrification, though pedestrian access to the other side of the track was maintained by the footbridge visible on the right. The station goods yard to its north on the southbound side survived into the British Railways era but is now a station car park.

Opposite below: **This view** taken looking south from Woodside Park Station's footbridge captures a seven-car train of Northern Line 1972 Mk1 Stock departing bound for Central London and beyond. It is of note how the platform canopy on the northbound platform is much shorter than that on the southbound, a feature common to other former GNR stations on this and the Enfield Town branch.

Above: **One of** London Transport's sturdy RF Class buses on route 251 took me across from the Northern Line's Totteridge & Whetstone Station to the Bakerloo Line's Stanmore terminus, where 1938 Stock driving motor 11076 awaits departure heading a seven-car train bound for Elephant & Castle. By this time, the '58 trailers' had all been withdrawn, the last having been scrapped in the summer of 1974. In the background may be seen the station's imposing terminus building, dating from the opening of the branch in 1932 by the Metropolitan Railway. The Bakerloo Line took over the branch in 1939, using newly-built tube tunnels between Baker Street and Finchley Road, with new intermediate deep-level stations replacing those on the Metropolitan Line between those points. Now, the branch was due to change hands again upon the opening of the Fleet Line, and on the right may be seen work progressing to renew the car sidings at Stanmore.

Opposite above: **The station** at Queensbury on the Stanmore branch was not added to it until December 1934 and was named thus as a 'companion' to the existing Kingsbury Station to its south. It actually gave its name to the surrounding area, which was rapidly built up with typical 1930s-style housing once the station was in use. This view sees 1938 Stock driving motor car 10316, another of the LNER batch, at the trailing end of a train departing from Queensbury for the south. This rearmost unit carries the recently introduced LT 'roundel' livery.

Opposite below: **A unique** station on the London Underground system was Drayton Park, the only open-air station on the old Great Northern & City Railway, at which trains call at the single island platform located between the tunnels heading southwards to Moorgate, and north to Finsbury Park. This view of its booking office on 20 April 1975 typifies its appearance when the line became attached (unfortunately not physically, owing to the abandonment of its extension over the GNR/LNER branch to Highgate and beyond) to the Northern Line. The station is little changed today.

THE LONDON UNDERGROUND, 1968–1985 • **109**

Above: **A small** depot existed at Drayton Park, adjacent to the northbound platform. On the same day as the previous picture, driving motor car 10115 heads one of two trains of 1938 Stock stabled there. These had replaced the last complete units of Standard Stock on the Underground system in the autumn of 1966. Many of them, however, saw further use on the Isle of Wight, surviving until as late as 1990.

Opposite above: **Following the** Northern City Line's 'decapitation' in the autumn of 1964, when it was truncated at Drayton Park in order for its terminal platforms at Finsbury Park to be used for cross-platform interchange between the Piccadilly Line and the new Victoria Line, only the northbound platform at Drayton Park was usually used for trains terminating and starting from there. The station saw little use, apart from on match days at nearby Arsenal Stadium, since passengers travelling to the northern suburbs from the branch changed to the Victoria Line at Highbury & Islington. The formation which 1938 Stock driving motor car 10113 heads here at Drayton Park is surely an over-provision for off-peak use.

Opposite below: **Had the** Northern Line's extensions over the LNER Northern Heights branches been completed in full, it would have been possible to travel directly from the Northern City Line, via Finsbury Park and Highgate, to East Finchley, but in the event anyone wanting to travel there, say, from Highbury had to use the 609 trolleybus, and its replacement the 104 bus. Today, route 263 provides the link. On 23 April 1975, I have used the bus to get there from home in Canonbury and find 1938 Stock driving motor car 10036 heading a through service from Morden to Mill Hill East via Bank.

Above: **On East** Finchley's southbound platform the same day, 1938 Stock driving motor car 10228 is at the trailing end of a service to Kennington via Charing Cross. Its guard looks forward along the platform to make sure all passengers have safely alighted or boarded before closing the train's doors. The large station car park on the left occupies the space where the goods yard had once been.

Opposite above: **On 8 May** 1975, 1938 Stock driving motor car 11161 gleams in the spring sunshine as it enters Finchley Central Station heading a train bound for Morden via Bank. It has come from the stunted Mill Hill East branch. As previously mentioned, the GNR Northern Heights branches split at Finchley Central, with the branch to Edgware (which was built first) heading straight on at the country end of the station, and that to High Barnet diverging sharply to the right. The junction for this may be seen towards the rear of the train. Had the 1935-1940 New Works Programme been completed in full, a new station booking office would have spanned the bridge seen here.

Opposite below: **On the** same occasion, sister 1938 Stock driving motor car 10234 is at the trailing end of a train that has come from Morden via Bank and is bound for Mill Hill East. It departs from Finchley Central's third platform which, over the years, has been used as a terminus for trains from Mill Hill East when the stunted branch has been served by a shuttle service between the two stations.

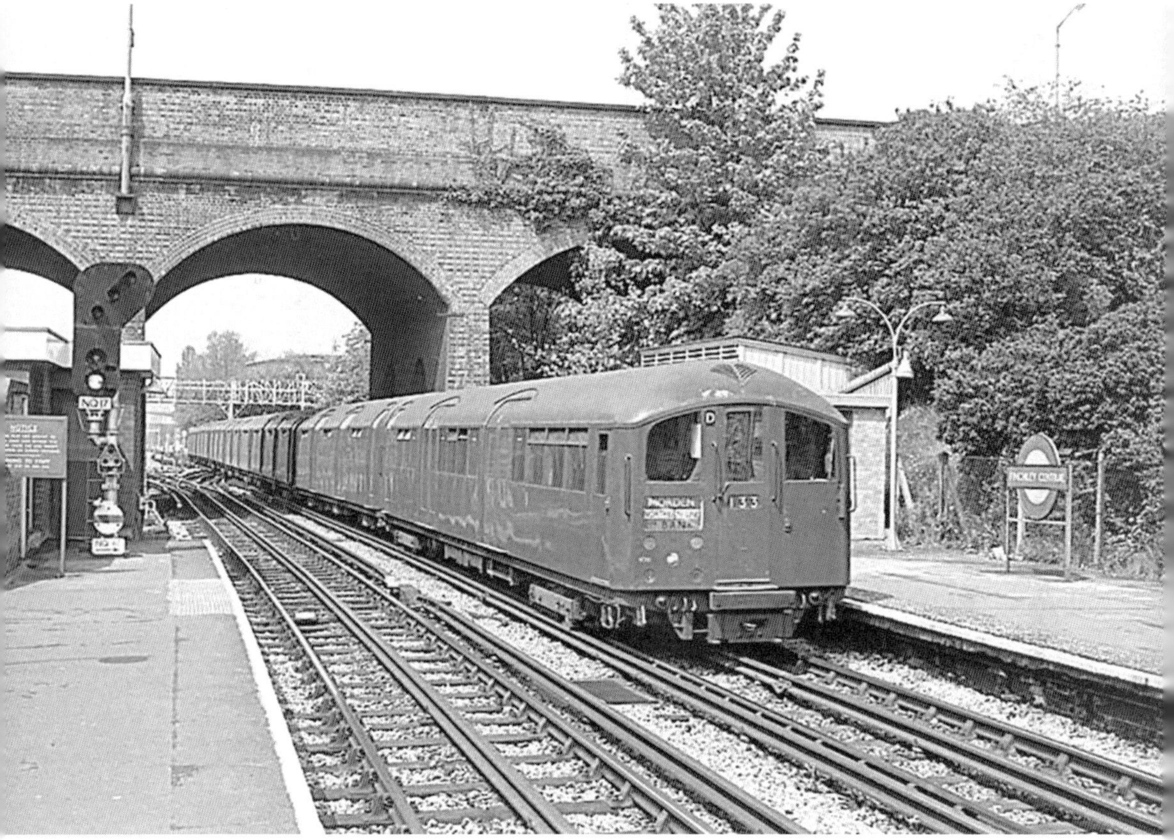
MORDEN
BANK
133

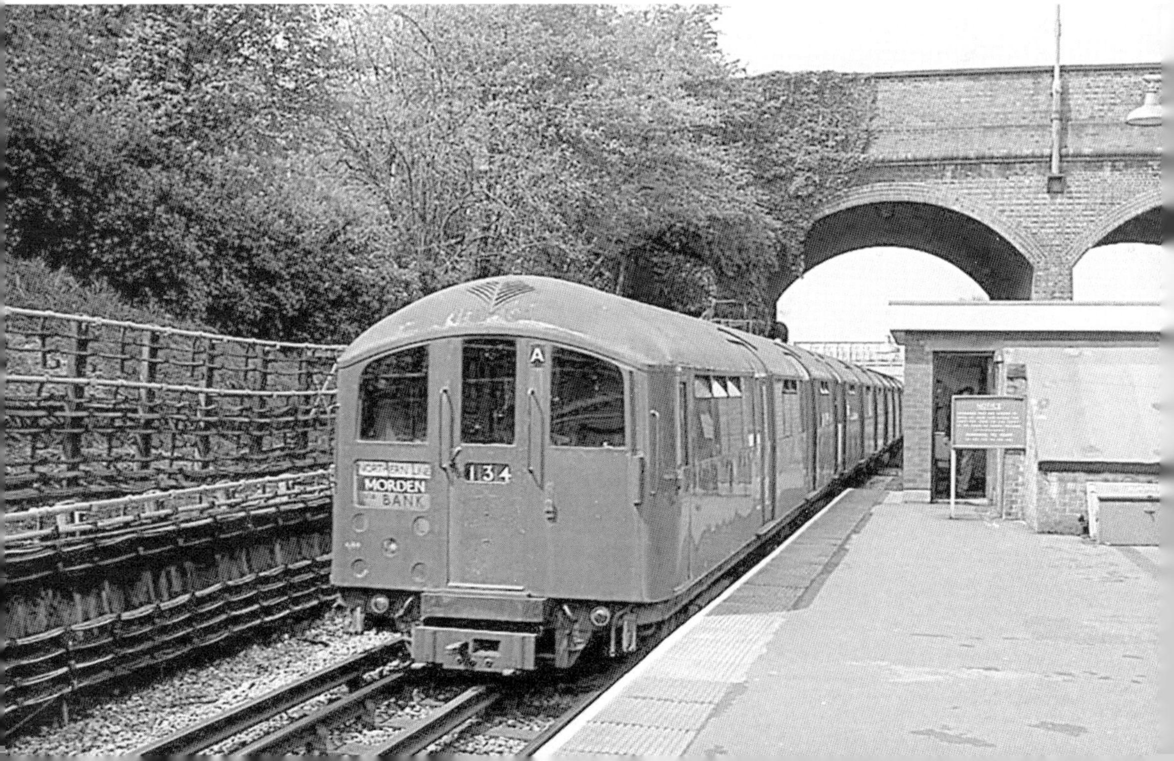
MORDEN
BANK
134

Above: **This view** of 1938 Stock driving motor car 10289, one of the LNER-owned cars, at the trailing end of the train from Mill Hill East, clearly shows how all of the original buildings at Finchley Central Station remain as they were prior to the line's electrification in 1940. On the extreme right is an antiquated waiting room for northbound passengers. This is the only structure that has been replaced in recent years.

Opposite above: **At Woodside** Park Station on the same day, as a seven car 1938 Stock train departs bound for High Barnet with driving motor car 11107 at its trailing end, two of the station's signal boxes are in evidence. The original Great Northern structure built in 1872 is nearest the camera, and once looked after the level crossing which was just beyond the platforms. A newer structure was provided further north, which controlled access to the station goods yard, now occupied by the car park. Both survived for many years after the line's electrification in 1940, being used to store materials, though the newer structure has been removed in more recent years.

Opposite below: **West Finchley** Station on the Northern Line's High Barnet branch also has some quaint old station buildings, but these are not in the same GNR style as its others. The reason is that the station was not built until 1932, and the LNER used structures removed from branch lines in rural East Anglia that had been closed. This is shown to good effect in this view of 'LNER' 1938 Stock driving motor car 10297 arriving there at the head of a train bound for Kennington via Charing Cross, also on 8 May 1975.

Above: **The spread** of suburbia in the early 1930s was the reason for West Finchley Station's appearance, and the station itself stimulated further new housing development in the area. 1938 Stock driving motor car 11209 leads a High Barnet-bound train into the station, with housing typical of the period behind it.

Opposite above: **Further to** the East in outer North London suburbia, the Piccadilly Line was extended north from Finsbury Park to Cockfosters in 1932/33, with all of the new stations being splendid Charles Holden designs typical of the era. Also on 8 May 1975, a train of 1959 Stock headed by driving motor car 1195 has arrived at the latter three-platform terminus, and its driver changes ends to prepare for the long journey to Rayners Lane, on one of the sections of the line extended westwards from Hammersmith at the same period. Holden-designed stations also grace that part of the line on the branches to Hounslow West and to Uxbridge, the terminus at the latter (on an existing Metropolitan Railway branch) being re-sited and rebuilt in very similar style to Cockfosters.

Opposite below: **Oakwood Station** is another Charles Holden design typical of the period, and on the same day, another 1959 Stock train with driving motor car 1228 trailing has called there, its penultimate stop before Cockfosters. The third track on the left leads to the southern end of Cockfosters depot, which is situated between the two stations, Oakwood having originally been named Enfield West.

THE LONDON UNDERGROUND, 1968–1985 • **117**

Above: **Arnos Grove** Station is another Holden design classic, and has three platforms, enabling trains to terminate there from the south when necessary. 1959 Stock driving motor car 1272 heads one such bound for Hounslow West, and still looks very smart after its April 1974 overhaul, over a year previously. These units had mostly been borrowed for the Central Line when new in the early 1960s and would soon be replaced by new 1973 Stock and transferred to the Northern Line. There are also stabling sidings to the south of Arnos Grove Station.

Opposite above: **At this** period, some off-peak trains terminated at Colindale Station on the Northern Line's Edgware branch. On 9 May 1975, 1938 Stock driving motor car 10160 is at the trailing end of one such service, and departs from the station heading for the reversing siding between the through tracks north of it. A train of 1972 Mk1 Stock arrives heading southbound on the right.

Opposite below: **Entering Colindale** Station on the same occasion is 1938 Stock driving motor car 10190, heading a train bound for Kennington via Charing Cross. The structure at the end of the platform is a relay room, housing signalling control equipment. Such buildings were erected, and equipped, on the uncompleted Northern Line extensions too, for example at Cranley Gardens on the Alexandra Palace branch.

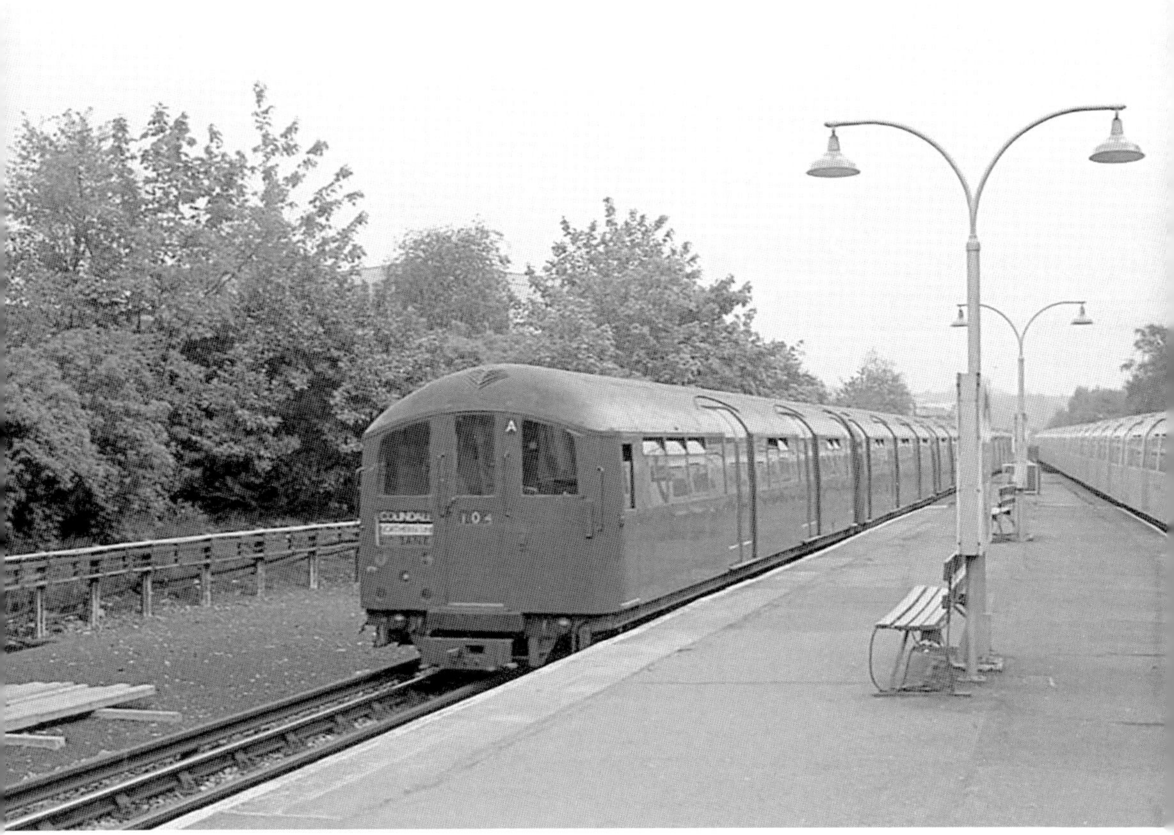

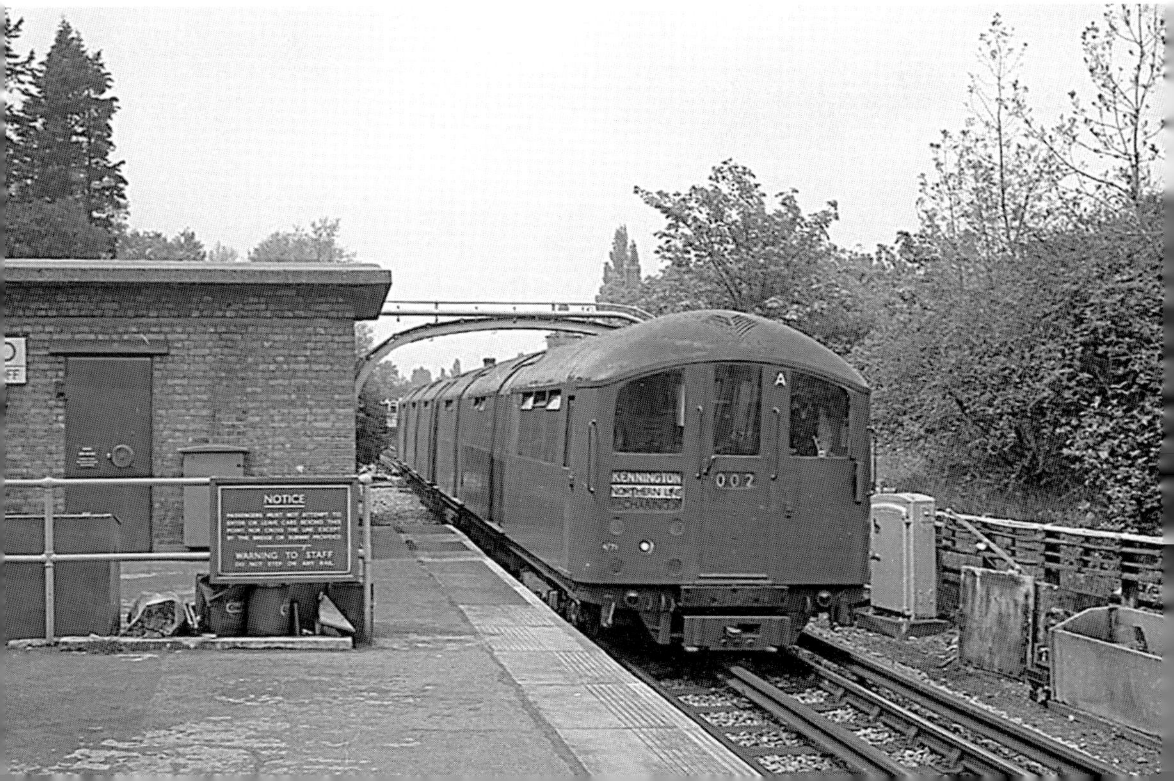
KENNINGTON
002
NOTICE
WARNING TO STAFF

Above: **Bound for** Stanmore, another train of 1938 Stock arrives at the more substantial, and slightly older, Kingsbury Station on the same occasion. Driving motor car 10330 is leading it. Just over four years later, this branch would be taken over by the Jubilee Line, using 1972 Mk2 Stock trains transferred from the Northern Line.

Opposite above: **Unlike neighbouring** Queensbury Station, Kingsbury has Metropolitan Railway architecture typical of the time of its opening in 1932. This is amply shown beside and above a 1938 Stock train arriving there bound for Elephant & Castle, with driving motor car 11020 leading.

Opposite below: **In contrast** to the pleasant suburban outer north and north-west London suburbs most 1938 Stock trains are serving on 9 May 1975, those on the East London Line are running through its murky tunnels and below the streets of Whitechapel, which Jack The Ripper terrorised in 1888. Very close to where the first ritual murder credited to him, in Bucks Row behind the original District Railway station, was committed, driving motor car 11106 leads a four-car set into the station on one of the rush hour workings from Shoreditch. Of note is the very deep step down from the platform to car floor level, something that would be strictly *verboten* today. Its destination 'display' would not satisfy today's regulations either.

Above: **At Shoreditch** itself, sister car 11098 is at the head of another four-car unit of 1938 Stock hopefully awaiting rush hour passengers. The difference in height here between the platform and the car floors appears to be not quite so severe. The second platform on the right has long been out of use; once trains had continued from here into Liverpool Street main line terminus.

Opposite above: **Correctly showing** its 'Shoreditch' destination plate, 1938 Stock driving motor car 10186 arrives at the terminus. Above the disused platform may be seen part of the former Great Eastern Railway Bishopsgate Goods Depot. This was closed after a disastrous fire in December 1964, but several of its features remained for many years afterwards, notably the wagon hoist visible at the top of this picture. The site was used mainly for car and lorry parking at this time.

Opposite below: **Owing to** the diabolical service record of the MB-type single-deck buses purchased by London Transport in 1967-69 to impose one-man operation on suburban bus routes, they were withdrawn after only six or seven years' service in most cases. Because few other operators wanted them, they soon ran out of garage space and had to store them on Underground premises before most eventually went for scrap. Three of these contraptions, with vehicles MBS253 and 259 nearest the camera, are dumped at the rear of the Victoria Line's Northumberland Park Depot on 7 June 1975. Its car sheds stretch southwards, and a train of two four-car 1967 Stock units stands in the open sidings. These are contemporary to the buses, and the bodywork of both modes of transport was built by Metro-Cammell in Birmingham. Unlike the short-lived MB-type buses, 1967 Stock continued to serve the Victoria Line until 2012.

Above: **The last** London Transport Underground trains ran on the Northern City Line on Saturday, 4 October 1975, latterly terminating at Old Street in order to allow work to proceed at their usual Moorgate terminus for its re-conversion for main line size rolling stock. 1938 Stock driving motor car 11115 is at the trailing end of a train that has terminated at Drayton Park that afternoon. The station is busier than usual because Arsenal are playing Manchester City; home fans would be disappointed, since they lost 2-3. A rail replacement service between Drayton Park and Moorgate was operated by Grey Green and National Express coaches until British Rail's Great Northern suburban electrification scheme was inaugurated in August 1976, finally linking the branch to the main line as had been originally intended, and also bringing into passenger use the ramps at Drayton Park built for the line's abortive extension over the Northern Heights branches.

Opposite above: **Typifying the** District Line at this period, a train of R Stock bound for Barking calls at Earls Court on 1 November 1975. R38 driving motor car 21125, of the 125 such cars converted from Q38 Stock trailers after the war, is at its trailing end.

Opposite below: **CO and CP** Stock comprised the rest of the District Line's rolling stock at this period, and on the same day, CP Stock driving motor car 53226 heads an Ealing Broadway-bound train at Barons Court Station. It contrasts with Piccadilly Line 1959 Stock driving motor car 1224 heading a Rayners Lane-bound train on the adjacent platform. At Barons Court, and all the way to Acton Town, District Line trains use the outer two tracks, whilst those on the Piccadilly Line, which emerge from their tunnel just east of Barons Court Station, use the inner two.

Above: **Car 1224** has taken me to Rayners Lane, where Piccadilly Line trains terminated at this period during off-peak times and at weekends, this being a Saturday. The train will stand in the reversing siding west of the station before returning to Central and North London. The typical Charles Holden architecture of the station booking office and platform buildings, dating from when the Piccadilly Line was extended here in 1932, provide a pleasant backdrop.

Opposite above: **Another train** of Piccadilly Line 1959 Stock, headed by driving motor car 1099, emerges from the siding at Rayners Lane to begin its long journey to Cockfosters. The signal cabin behind it is also typical of Underground architecture from the time of the Piccadilly Line 1932/33 extensions at both ends of the line.

Opposite below: **On 27 April** 1979, four days prior to the Jubilee Line opening, a train of 1972 Mk2 Stock heads west of Willesden Junction Station on the Bakerloo Line. Such units had been transferred to it from the Northern Line, upon that line's receipt of 1956/1959 Stock from the Piccadilly Line when it was displaced by new 1973 Stock.

Above: **All looks** spotlessly clean at the new Jubilee Line Charing Cross Station on 7 May 1979, six days after opening. Driving motor car 3259 is at the trailing end of a 1972 Mk2 Stock train which has arrived there. All such units were transferred from the Bakerloo Line to the new line, although the 1972 Mk1 units remained on the Northern Line. Just over twenty years later, all of the latter in their original formations were withdrawn, the section of Jubilee Line from Green Park to Charing Cross was closed, and the 1972 Mk2 units were back on the Bakerloo Line – where they still are at the time of writing, and are likely to stay for a long time yet.

Opposite above: **The much-depleted** Bakerloo Line now reverted to entirely 1938 Stock operation, the only tube line still operating red trains. On 15 May 1979, driving motor car 10258 (one of the LNER-owned cars) heads one of the few remaining rush hour trains to Watford Junction into Willesden Junction 'New' Station. The high level part of the station on the North London Line is on the bridge above it.

Opposite below: **The only** other line on the Underground operating red trains by now was, of course, the District whose CP Stock driving motor car 53228 has arrived heading a train of CO/CP Stock at Richmond on 16 May 1979. It disgorges rush hour commuters, and contrasts with a British Rail London Midland Region North London Line Class 501 EMU in the adjacent platform.

Above: **CO Stock** driving motor car 53010 has also arrived at Richmond with a similar formation on the District Line and has recently been overhauled in the new livery with smaller white fleet numbers and LT roundels instead of fleet names. Its leading end is framed by ornate LSWR ironwork dating back to when the station was built.

Opposite above: **By this** period, the unpainted aluminium silver-liveried 1956, 1959 and 1962 Stock trains which looked so shiny and sleek when new were beginning to show signs of age, and even before they were befouled by the curse of graffiti, looked more like a dirty grey than their original silver. 1962 Stock driving motor car 1624 heads a four-car unit which has arrived at the Central Line's remote Ongar terminus on 17 June 1979. By now, the original Great Eastern Railway signal box at the London end of the platform has been taken out of use and is boarded up, the station goods yard having long since been disused.

THE LONDON UNDERGROUND, 1968–1985 • **133**

Below: **A peep** through the fence of Highgate Wood Sidings on 1 July 1979 finds battery locomotive L36, one of the first nine of this general appearance which were supplied to LT in 1936/37, with some flat wagons. It has been used to power an engineer's train when the current is switched off during non-traffic hours. I took this picture from the trackbed of the branch from Highgate to Alexandra Palace which should, of course, have become part of the Northern Line complex in 1940/41.

Above: **On the** same occasion, Highgate High Level Station, eerily situated in a deep cutting between Highgate East and West Tunnels on the former GNR/LNER Northern Heights branch from Finsbury Park to Alexandra Palace, Edgware and High Barnet, is gradually being engulfed in foliage. Apart from its toilet blocks at the London end of the island platform, the station was completely rebuilt in 1939/40 for the extension of the Northern City Line over the branch but, of course, never served by tube trains and closed to passengers when the service to Alexandra Palace was withdrawn by British Railways in July 1954, after the extensions had been abandoned. The new structure was built in connection with the new tube station directly below it on the extension of the Northern Line from Archway to High Barnet and Edgware via Mill Hill East, two short flights of steps lead from its platforms to the booking hall and top of the escalators to the deep-level platforms below. Water seepage from the disused trackbed causes problems here, but the disused station was integrally built with the lower part of the intended tube station interchange, so its demolition would cause difficulties. It therefore remains to this day as a haunting reminder of the wasted works on the uncompleted Northern Line extensions.

Opposite above: **The former** GNR line's bridge leading into their Edgware Station, from which I took this view of 1959 Stock driving motor car 1084 heading a Northern Line train bound for Kennington via Charing Cross, is also still in place on 1 July 1979. A train of 1972 Mk1 Stock and others of 1956/59 Stock are visible in the sidings behind it, adjacent to the LT Edgware terminus.

Below: **Not far** away from the previous picture, two trains of Jubilee Line 1972 Mk2 Stock gleam in the autumn sunshine in the sidings at Stanmore Station on 16 September 1979. Two others are visible at the station's platforms on the right. Around this period, London Transport examined proposals to extend the line from here the short distance to their Aldenham Bus Overhaul Works, where space was available to establish facilities for maintenance of Underground rolling stock to replace the antiquated Acton Works. Since the Stanmore branch had been built to main line loading gauge, stock from the entire Underground system could have been dealt with there, and Aldenham would have fulfilled its original purpose, since of course it was originally built for the Northern Line's new extension from Edgware to Bushey Heath. But it was not to be, and the bus overhaul works was closed in 1986 as a result of the Thatcher regime's anti-public transport policies.

15 December 1979, a dismal winter's day, finds R38 Stock driving motor car 21113 approaching the District Line's Becontree Station heading a seven-car formation of R Stock bound for Richmond. The vast inter-war London County Council Becontree Estate, once the largest municipal housing estate in Europe, stretches into the distance.

By Sunday 23 March 1980, the Epping to Ongar shuttle on the Central Line was again under threat of closure. On that day, the usual four-car set of 1962 Stock, on this occasion led by driving motor car 1577, arrives at Blake Hall Station, little-used even in rush hours.

With open farmland to the left, 1962 Stock driving motor car 1576 returns to Blake Hall Station heading the same train on its way to Ongar. Could the number of people on the platform present a record loading for this quiet station? No. They are enthusiasts on a North London Transport Society tour exploring this remote area's former bus routes on a preserved ex-London Transport RF. Not long after our visit, the Epping to Ongar shuttle was reduced to operate on Monday to Friday rush hours only, and at the end of October 1981, Blake Hall Station was closed.

By the time of our visit to this remote Essex outpost of the London Underground, the rush hour short workings to and from North Weald Station had been withdrawn, and the second track used to accommodate them lifted. This view looking south at the station shows the second platform built by London Transport for these workings, along with a very rudimentary waiting room. The original Great Eastern Railway signal box has, fortunately as it was to turn out, been boarded up to protect it from vandalism.

Above: **In total** contrast to the previous three pictures, Wembley Park is a very busy station, especially when sporting events are in progress at the nearby stadium. It is served by the Jubilee and Metropolitan lines, and on 10 May 1980 an eight-car train working a fast service to Amersham on the latter, headed by A62 driving motor car 5223, approaches the station. The non-electrified tracks in the foreground are those of the former Great Central Railway, whose main line north of Aylesbury was axed under British Railways' notorious 'Beeching cuts' in 1966.

Opposite above: **London Transport** organised visits to their installations, as well as enthusiasts' societies, and on 13 May 1980 I was able to secure a visit to the Underground's Acton Works on behalf of the North London Transport Society. This view finds an assortment of car bodies, resting on 'dummy' bogies, awaiting overhaul having been dismounted from their own. They have been placed here by means of a traverser. Unfortunately, the only car I can positively identify is District Line R49 Stock non-driving motor car 23572 in the centre. It is flanked by tube stock of either 1959 or 1962 Stock variety, with tube cars of either 1967 or 1972 Mk1 stock to the extreme left and right.

Opposite below: **Central Line** 1962 Stock non-driving motor car 9571 is on the traverser, waiting to be moved to its relevant place during the overhaul process. Some of these were built out of LT numerical sequence by British Railways in 1961, to provide an eighth car in the seven-car formations of 1959 Stock borrowed from the Piccadilly Line, to expedite removal of ageing Standard Stock trains from the Central Line.

Above: **A stage** further in the overhaul process, overhauled car bodies await reunification with their running units. On the left is a car of 1967 or 1972 Mk1 tube stock, next there is a 1938 Stock tube car. In the centre is a C69 Stock car which has had its first overhaul, with another 1967/1972 Mk1 and a 1959/1962 Stock car on the right. This view amply exemplifies the difference in height between tube stock and sub-surface stock on the Underground.

Opposite above: **Inside the** Lifting Shop at Acton Works, where car bodies are dismounted from their bogies, etc., C69, 1959/62, 1938 and 1967/72 Mk1 cars are in evidence. At this period, when ageing 1938 Stock was keeping the Bakerloo Line going, these trains were being given extra heavy overhauls here, and were referred to as 'EHO' cars.

Opposite below: **An oddity** at Acton Works on this occasion is car G663, a former Standard Stock car known as the 'rail gauging car'. Its purpose was to make sure that there was sufficient clearance in running tunnels, for example where engineering works had been carried out renewing tunnel segments on tube lines.

Above: **What looks** like a station platform is in fact one of the reception bays at Acton Works. Stabled there awaiting its place in the works is a train of Piccadilly Line 1959 Stock, with driving motor car 1181 facing the camera.

Opposite above: **In the** adjacent reception siding at Acton Works is a similar train of Central Line 1962 Stock, with driving motor car 1519 facing me. At first sight identical to the 1959 Stock in the previous picture, these units had minor differences from them, one notable visual difference being that the red tail lamp beneath the offside cab window was on the left, rather than the right of the twin headlamps. However, the two stocks could, and did, work together in multiple formations. As related earlier, the 1962 Stock was in effect a follow-on order from the 1959 units, both types being built by Metro-Cammell with the exception of the B.R.-built 1962 Stock non-driving motor cars, to expedite replacement of old stock on the Central Line.

Opposite below: **A last** look at Acton Works on 13 May 1980 shows Metropolitan Line A62 Stock driving motor car 5218 heading its four-car unit after overhaul in the works' output sidings. As may be observed, the cab area looks far from its sleek, silver original self after some eighteen years or so service. This was because the aluminium body exteriors, notably the cab's centre door, became pitted with stones and so on during the trains' journeys. The body sides are still presentable though, yet to be defaced by the childish graffiti which would plague the Underground a few years later. Both A60 and A62 units' bodies were constructed by Cravens of Sheffield, which must have compensated the firm for the 1960 Stock, whose prototype driving motor cars they bodied, not being adopted as the replacement for the Central Line's Standard Stock.

THE LONDON UNDERGROUND, 1968–1985 • **143**

Above: **By this** time, R Stock cars which had originally been red, and then painted silver in the 1950s and early 1960s, were painted white. One of these, R38 driving motor car 21119, heads a Richmond-bound train heading west from Turnham Green Station to join the North London Line's Richmond section at Gunnersbury. The date is also 13 May 1980.

Opposite above: **Deliveries of** new District Line D78 Stock had, by 31 March 1981, ousted virtually all surviving CO/CP Stock on the line, with R Stock their next victims. On that day, 22656, another R38 driving motor car, heads an Upminster-bound service into West Kensington Station.

Opposite below: **Driving motor** car 21104, another Q38, is at the trailing end of the same train as it departs for the East from West Kensington Station. An obvious difference from R Stock cars that were rebuilt after the war from Q38s was that whereas these had four windows between their two sets of double passenger doors, the newer cars (the majority being R47, R49 or R59 non-driving motor cars) had two larger windows instead. This may clearly be observed here with the two cars beyond 21104.

THE LONDON UNDERGROUND, 1968–1985 • 145

Heading east from Ravenscourt Park Station, R38 Stock driving motor car 21130 is at the trailing end of a service terminating at Dagenham East, rather than continuing to the District Line's easternmost terminus at Upminster. I took the picture from the westbound platform; the two tracks in the foreground are the Piccadilly Line fast tracks.

Also at Ravenscourt Park Station that day, 22678 is another R38 at the trailing end of an Ealing Broadway-bound service. The reason I was in this area on 31 March 1981 was to photograph and ride on the last CO/CP Stock train in regular public service.

THE LONDON UNDERGROUND, 1968–1985 • 147

The train in question, led by CP Stock driving motor car 53267, took me from Ravenscourt Park Station to Richmond, where it has terminated. A nicely made special headboard reading 'Last red train on the District Line', produced by some of the line's operating staff, is displayed.

Another CP Stock driving motor car, 54231, is at the eastern end of the train, and in common with the rest of its cars, is one of those that gained the new 'roundel' livery. A clutch of other railway enthusiasts gather at the end of the platform to record this historic occasion on film. In the distance, a number of former London Transport RT-type buses are visible in the yard of independent bus and coach operator Continental Pioneer. A dealership specialising in selling such vehicles for preservation shared the yard.

Above: **The train** was bound for Dagenham East, where 54231 has led it into the bay platform to terminate. As may be observed, this car has an additional notice 'Farewell CO/CP, 1938 - 1981' in its cab window. A new D Stock train arrives on the adjacent platform, bound for Upminster.

Opposite above: **Most of** the passengers on this last red train on not only the District Line but also all the London Underground's sub-surface lines were enthusiasts, and we throng the platform at Dagenham East for a return trip to Richmond led by car 53267. It took us beneath the City and West End at the height of the evening rush hour, thus ensuring quite a lively trip.

Opposite below: **Just under** three weeks later, on 19 April 1981, a Farewell Tour for the CO/CP Stock ran. Whist waiting to apprehend it with my camera, I again took the opportunity to photograph R Stock, now themselves being rapidly withdrawn. R38 Stock driving motor car 21100, numerically the first one, brings a Richmond-bound train into Plaistow Station. Of note is the ornate ironwork showing the initials of the station's builder, the London, Tilbury & Southend Railway. Also of note is the bay platform on the left, at which some District and Hammersmith & City Line trains terminated, particularly if curtailed owing to late running.

THE LONDON UNDERGROUND, 1968–1985 • **149**

150 • THE LONDON UNDERGROUND, 1968–1985

Above: **Closer to** Central London, where the District Line comes briefly into daylight at Whitechapel Station, 21133, another R38 driving motor car, is at the trailing end of an Upminster-bound train that is heading back into the tunnel, from which it will emerge east of Bow Road Station. Below the trailing car are the East London Line's tracks and two platforms. As related earlier, Whitechapel Station is much changed today.

Opposite above: **On this** occasion, I did not ride on the tour train, but as I had done with the District Centenarian tour in 1969, I had its times and was able to get ahead of it at various points. I have done so at Wembley Park Station, where P Stock had originally worked, and did so on Metropolitan Line Uxbridge services until replaced by A Stock in the early 1960s. In this view, CP Stock driving motor car 53249 is in the lead. The headboard is in District Line green, with white LT roundel.

Opposite below: **For a** photo stop, the tour train paused for a while at Wembley Park, enabling me to get ahead of it once more to catch this view of it passing through West Harrow Station. Car 53249 is leading, evidently with a cab full of people. Once again, such units served this station when working Metropolitan Line services.

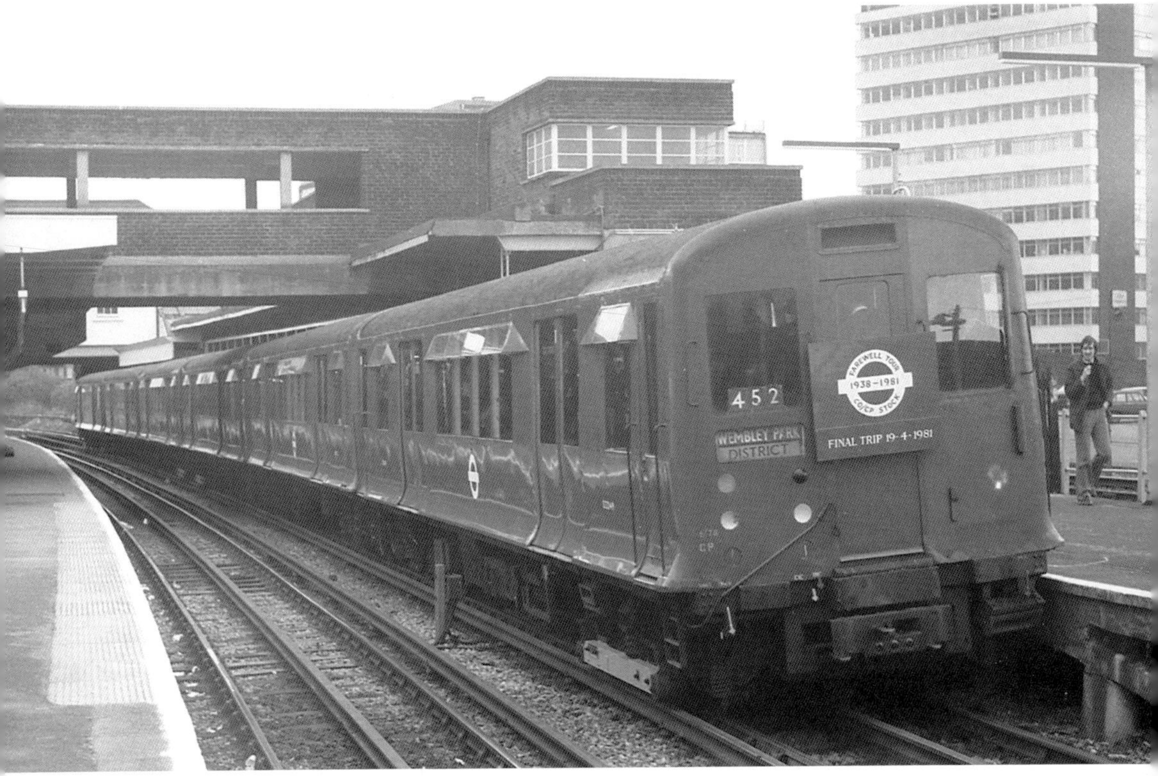
452
WEMBLEY PARK
DISTRICT
FAREWELL TOUR
1938–1981
CO/CP STOCK
FINAL TRIP 19-4-1981

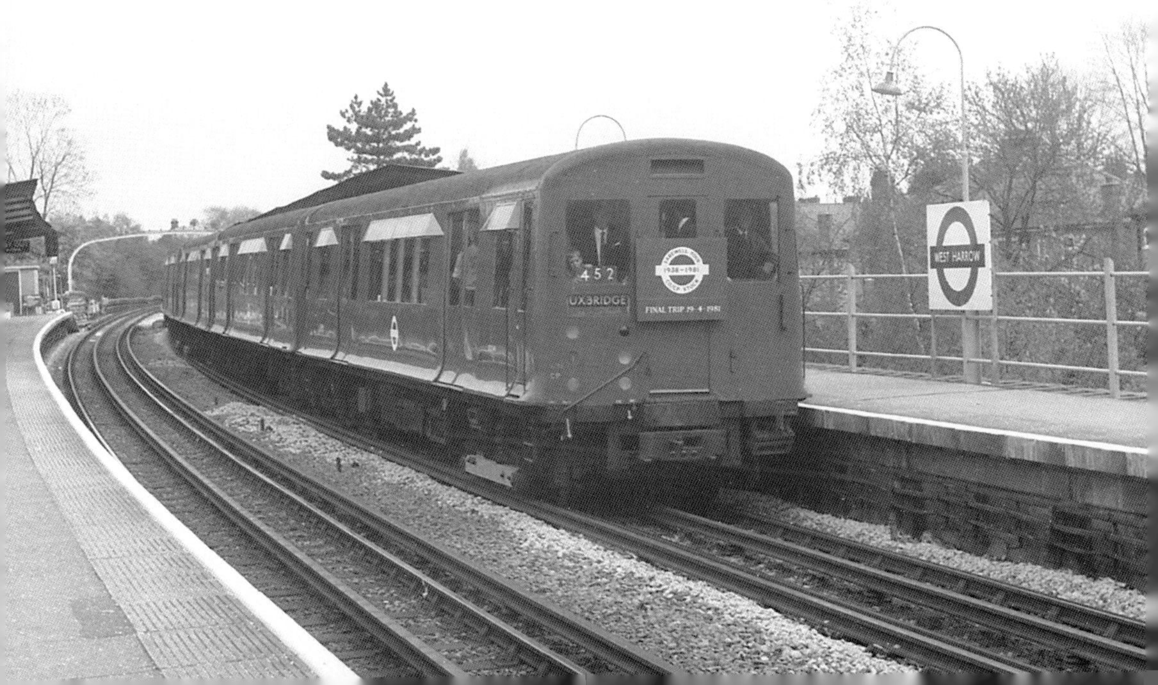
452
UXBRIDGE
1938–1981
FINAL TRIP 19-4-1981
WEST HARROW

Above: **The tour** stopped at Uxbridge for lunch, thus I was able to get ahead of it again. The next section of it took it along the Piccadilly Line, travelling through former District Railway territory to regain the present-day District Line at Ealing Common. Here, with CP Stock driving motor car 54265 leading, it passes through Park Royal Station, which as the station buildings on the left show, was rebuilt in splendid Charles Holden style for the Piccadilly's extension over the branch in 1932.

Opposite above: **On the** same day, a train of District Line R Stock departs from Earls Court Station for Ealing Broadway. Of note is the 'West Brompton Closed' sign on the wall on the left reminding motormen not to stop there, as it is a Sunday. Today, that station is an important interchange between the District Line's Wimbledon branch and the London Overground and Southern National Rail services on the West London Line and is of course open daily.

Opposite below: **The CO/CP** Farewell Tour, meanwhile, made its way via Earls Court and a reversal at Edgware Road District/Circle and H&C Line Station to Hammersmith Metropolitan Line Station. Here, with CP Stock driving motor car 54265 ready to head east again, an interesting contrast is made with C77 Stock driving motor car 5711, working a Hammersmith & City Line service to Whitechapel.

THE LONDON UNDERGROUND, 1968–1985 • **153**

Above: **At the** buffer stops at Hammersmith Met., the other ends of each train are CP Stock driving motor car 53249, and C69 Stock driving motor car 5577. The C69 Stock originally replaced CO/CP Stock on the Circle and Hammersmith & City lines, and C77 Stock was basically a run-on order of the same type, initially needed because there were clearance problems working new District Line D78 Stock between Edgware Road and Earls Court on Wimbledon services. As these pictures show, C69 and C77 Stock worked together in mixed formations of the two types. They could be easily distinguished by the earlier cars having black roofs, the later ones, silver.

Opposite above: **The Farewell** Tour ended at Ealing Broadway, before running back (officially) empty to Ealing Common Depot. An everyday appearance at the station there then is that of Central Line 1962 Stock, of which driving motor car 1596 has arrived heading a service which will return to Hainault via Newbury Park. It still looks smart, with no graffiti in evidence. This too, of course, is now an historic picture.

Opposite below: **Once the** District Line's CO/CP Stock had been withdrawn, mass withdrawal of its R Stock followed as deliveries of new D78 Stock arrived and entered service to replace them. Many were scrapped by C.F. Booth of Rotherham, where R49 Stock non-driving motor car 23571 still looks good for several years' more service – and probably was. Redundant London buses were scrapped there too – one of the last London Country RTs, RT4563, awaits its fate on the right. The date is 5 October 1981.

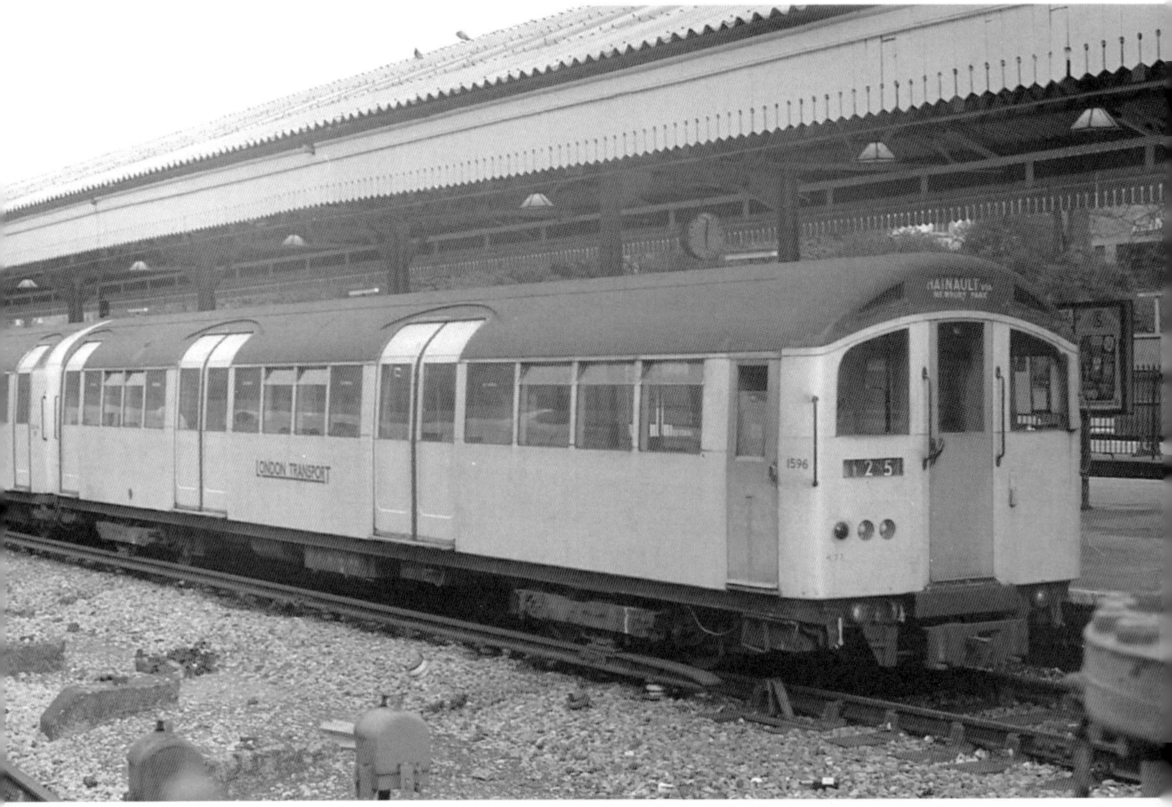
LONDON TRANSPORT
1596
125

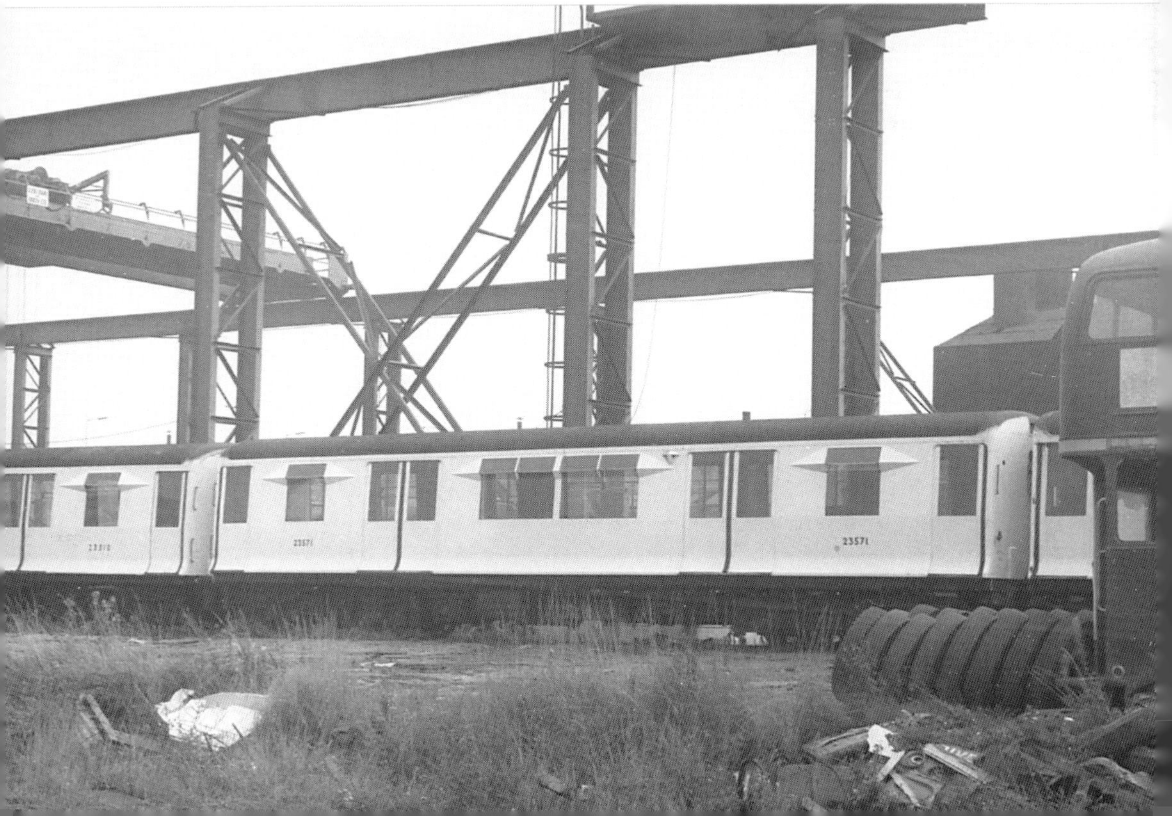
23571

Above: **At the** end of the scrapyard siding, R38 Stock driving motor car 21102 also awaits cutting up. Whereas this type, converted from Q38 Stock trailers, had steel bodywork, the R49 non-driving motor cars were of lightweight aluminium construction. The writing on the cab front of 21102 tells the scrap man to return its coupler for reuse to London Transport.

Opposite above: **By 22 November** 1981, such had been the problems caused by water seepage from the disused Highgate High Level Station to the booking office and escalator shaft of the tube station below, that all foliage in the affected area has been cleared. Of note in this picture, looking north, is the building on the left meant to house up and down escalators from Archway Road, where there would have been a new booking office (topped by a statue of Dick Whittington) serving both parts of the station. The latter was never built, and after several years of disuse, the space intended for the escalators was partially used by the provision of an up-only escalator, with a small structure at the top serving as an exit only into Archway Road.

Opposite below: **Another relic** on the London Underground system at this period is the entrance to the former Central London Railway/Central Line Wood Lane Station. This had been closed in 1947, when the new station at White City nearby replaced it. Seen on 28 November 1981, the structure was later demolished, although part of it was saved and reassembled at the LT Museum's Acton Depot premises.

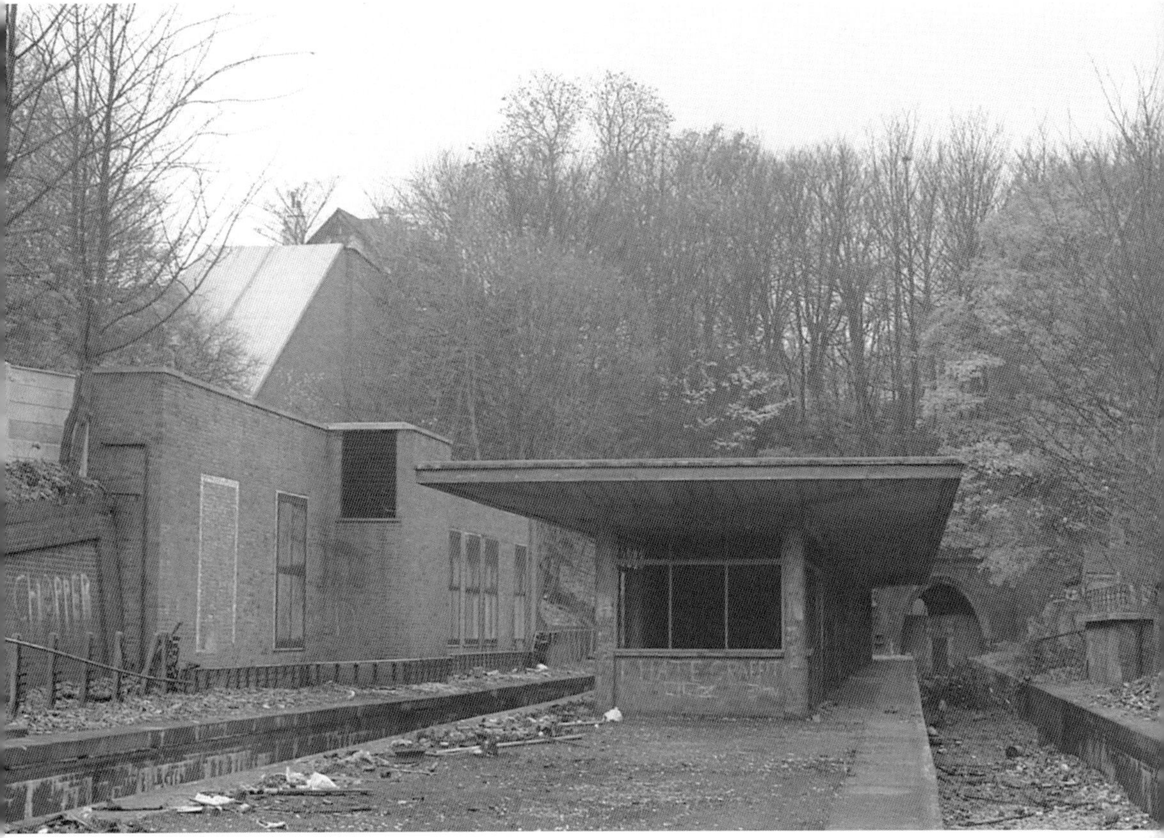
CHOPPER

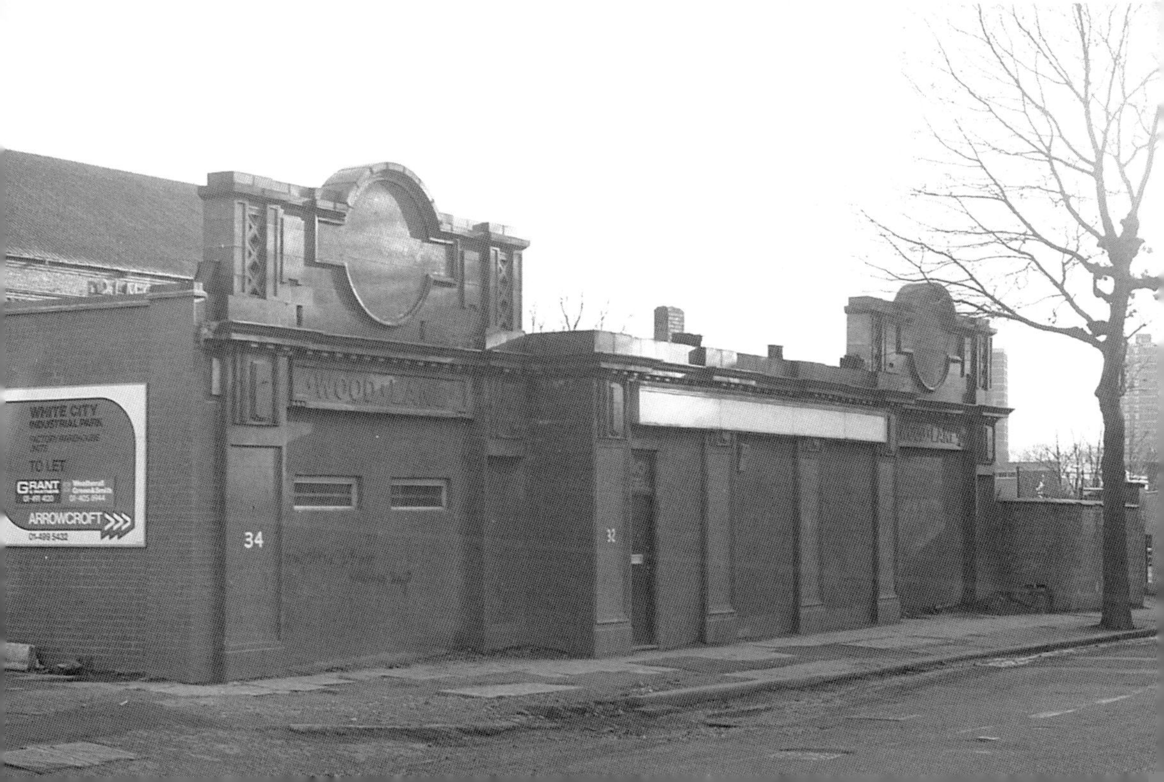
WHITE CITY
INDUSTRIAL PARK
TO LET
GRANT
ARROWCROFT
34
32

Above: **Three forms** of rail transport are evident in this view of the country end of Willesden Junction 'new' station on 17 June 1982. A car transporter train makes its way from the North London Line to the West Coast Main Line on the left, whilst a train of 'EHO' 1938 Stock, with driving motor car 10308 – one of the LNER-owned ones – trailing, arrives at the station bound for Elephant & Castle on the Bakerloo Line. On the right, a London Midland Region class 501 F M.U. works a Watford-bound local service.

Opposite above: **At this** period, thanks to the mutual loathing between Greater London Council Leader Ken Livingstone (who was ultimately responsible for London Transport policy) and Tory Prime Minister Margaret Thatcher, London Transport was being made to make massive cuts to bus and Underground services following its popular cheap fares policy being ruled illegal. One cut was the withdrawal of the few remaining rush hour journeys from and to Watford Junction on the Bakerloo Line. 17 June 1982 was planned to be their last day, and here another train of 'EHO' 1938 Stock, headed by driving motor car 10231, approaches Wembley Park Station on the first of the four evening Watford journeys that day.

Below: **1938 Stock driving** motor car 10275, another of the LNER-owned examples, arrives at Harrow and Wealdstone Station with the second of the four Watford Junction journeys that evening. Just ahead of where I took this picture is where the dreadful multiple collision occurred on the West Coast Main Line on 8 October 1952, when an overnight express from Scotland ploughed into the back of a stationary commuter train collecting passengers at the height of the morning rush hour. Seconds later, a double-headed express from Euston to Liverpool and Manchester ploughed into the wreckage, its two locomotives 'leapfrogging' the derailed engine of the up express, mounting the country end of the platform on the left and landing on the southbound Bakerloo and Watford local electric line. Had their doing so not cut off the current, a train of 1938 Stock approaching the station from the north may also have been involved in the disaster.

Above: **The third** Bakerloo Line Watford Junction train, with 1938 Stock driving motor car 11180 at its trailing end, calls at Headstone Lane Station. Of note is how the destination plate shows 'Watford (LMR)', to distinguish its terminus from the Metropolitan Line one elsewhere in the town. For many years after nationalisation, these plates still showed 'Watford (LMS)'.

Opposite below: **What is** intended to be the last Bakerloo Line train of all to Watford Junction, complete with an ornate sign in the cab window saying so, is headed by 1938 Stock driving motor car 10168, which calls at Bushey Station and will take me the two further stops to its terminus on 17 June 1982.

Right: **Upon arrival** at Watford Junction, car 10168 is dwarfed by an LMR Class 501 working the local Broad Street and Euston services to Watford. Many transport enthusiasts had ridden on the Bakerloo Line train, and many others were present on the platform to photograph this last train before the service was withdrawn – but it never happened after all.

Below: **Evidently because** the correct procedures had not been followed to allow withdrawal of the Bakerloo Line's Watford Junction service, it continued for another three months and one week. Thus, on 23 September 1982, the last train really did run, and as dusk falls, it arrives at Headstone Lane Station with former LNER-owned 1938 Stock driving motor car 10255 leading. It has a headboard proclaiming the event, in the same style as that used in June and also has the by now disused headlight destination indicator in use. However, the code it is showing was used for Wembley Park, not Watford Junction.

Above: **Less than** two years after the last CO CP Stock ran, final withdrawal also befell the District Line's R Stock trains. The last of all in normal public service is headed by R38 Stock driving motor car 22674 as it arrives at Stamford Brook Station on 4 March 1983. District Line staff have produced commemorative notices in the windows of each end cab.

Opposite above: **The trip** aboard this last R Stock train in normal service took me to Tower Hill, where it has terminated in the evening rush hour. A well-filled train of D78 Stock passes on the adjacent platform. These would provide the majority of the District Line's services for the next thirty years or so.

Opposite below: **With fellow** Q38 driving motor car 21121 leading, the train then ran from Tower Hill to Richmond. From there, it travelled to High Street, Kensington where it has arrived in one of its terminal platforms in this view, showing the correct headlight code. After this, only A Stock, C69/77 and D78 Stock ran on the Underground's sub-surface lines, until the new S Stock replaced all of them between 2010 and 2017.

THE LONDON UNDERGROUND, 1968–1985 • 163

Above: **At a** time when London Transport's very existence was being imperilled by the feud between its GLC masters and the Thatcher regime, it still found time to organise special events to celebrate its golden jubilee in 1983. On 23 April that year, an open day was held at Neasden Depot, at which a train of 1938 Stock from the Bakerloo Line, by now withdrawn north of Stonebridge Park, was displayed. Driving motor car 11156 heads it in this view.

Opposite above: **Amongst a** number of the Underground's non-passenger rolling stock on display at Neasden that day is battery locomotive L32, one of eleven supplied by Metro-Cammell in 1965. By now, such service locomotives are now in a new yellow livery, rather than the previous maroon. Of note is how L32 has its massive buffers, used when hauling main line-size stock, folded down.

Opposite below: **Nicely cleaned** up following its use a few weeks earlier as the last District Line R Stock train in service, and then on a Farewell Tour which unfortunately I was unable to attend, the formation is proudly displayed at the Neasden open day. R38 driving motor car 21121 is nearest the camera. Unfortunately, it did not survive into preservation.

THE LONDON UNDERGROUND, 1968–1985 • **165**

Above: **Such were** the cuts London Transport were forced to make to Underground services following the 'Law Lords' debacle regarding their cheap fares policy, that a number of 1959 Stock units became surplus to requirements on the Northern Line. These were therefore transferred to the Bakerloo Line to allow replacement of some of their 1938 Stock units. This view of driving motor cars 10151 (1938) and 1244 (1959) at Queen's Park Station clearly shows the similarity of design between them. Their internal layouts were similar too, except that on the 1956, 1959 and 1962 Stock the transverse seats in the centre of each car were arranged in two pairs of two facing each other, rather than one pair of two facing each other, each with a double seat behind them facing the same way. The date is 3 June 1983; almost exactly a year later, Bakerloo Line services were reinstated as far north as Harrow & Wealdstone, which remains the line's northern terminus today.

Opposite above: **Also at** Queen's Park Station on 3 June 1983, 1938 Stock driving motor car 10167 heads a Bakerloo Line service bound for Stonebridge Park, where trains terminated and then stood in the depot to the north of the station before returning south. However, it is having a crew change here and the driver has his cab centre door open.

Opposite below: **Kensal Green** Bakerloo Line Station is shared with the British Rail local Broad Street and Euston to Watford services, as are all those north of this point. 1959 Stock driving motor car 1224 heads a Stonebridge Park bound service in this view taken from the station footbridge as a West Coast Main Line down express speeds by.

Above: **Reductions in** requirements for Metropolitan Line services meant that some A Stock units were available to work the East London Line as four-car trains. On 23 June 1983, A60 driving motor car 5059 heads one of these at its New Cross terminus. Unlike that at New Cross Gate, this is not a through platform that can connect with the main line, but a shorter bay platform than the station's others. On the down one of these, Bulleid-designed 4EPB unit 5155 is standing, too.

Opposite above: **On the** first weekend of July 1983, exactly fifty years after the London Passenger Transport Board commenced its unification of our public transport systems, a two-day event threw LT's adjacent Chiswick Bus Overhaul Works and Acton Underground Works open to the public, spectators being able to walk through from one to the other thanks to a bridge beneath the North London Line which separated them. On 2 July, former Standard Stock driving motor car L130 is one of two (the other is L131) of the later 1934-built variety converted from cars 3690 and 3693 in 1967 to become pilot motor cars, used for shunting and so on. They carry the new yellow livery replacing the previous maroon. L130 has since been preserved by the LT Museum.

Opposite below: **By this** time, withdrawn 1938 Stock cars were also being used for non-passenger carrying duties, generally replacing older Standard Stock in such use. A more exotic conversion, however, is that of two driving motor cars, with three purpose-built trailers between them, to become the Underground's tunnel cleaning train, the cars being numbered TCC1-TCC5. With its yellow livery, it gained the nickname 'the big yellow duster' and is also on show at Acton Works. Car TCC1, rebuilt from driving motor car 10226, is nearest the camera.

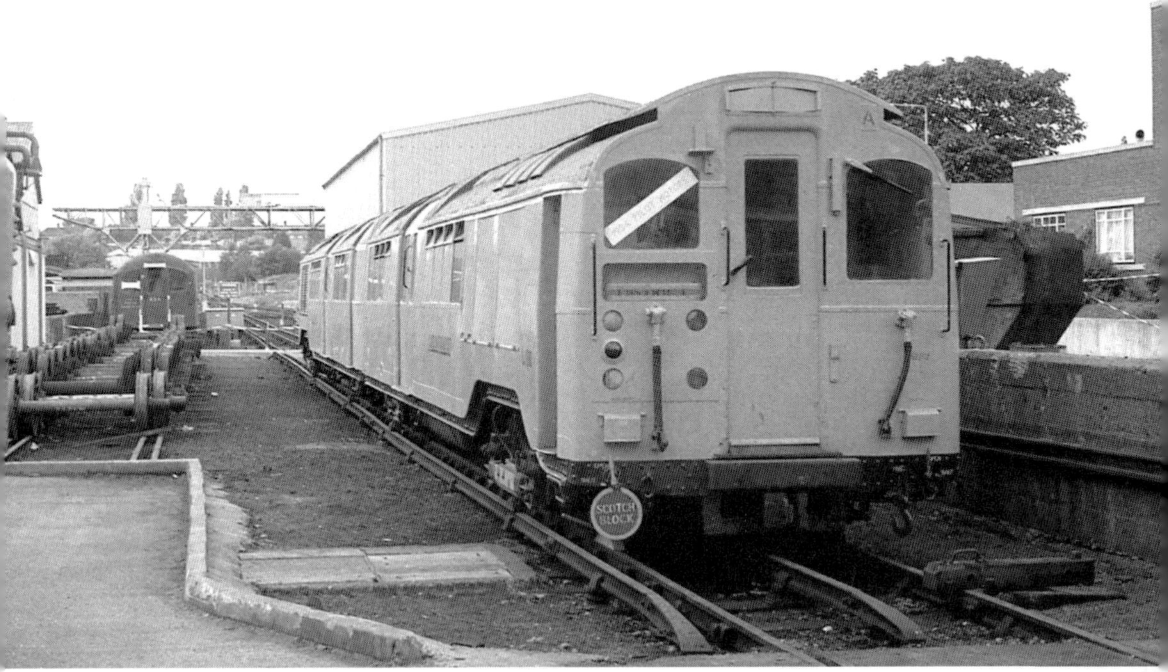
SCOTCH
BLOCK

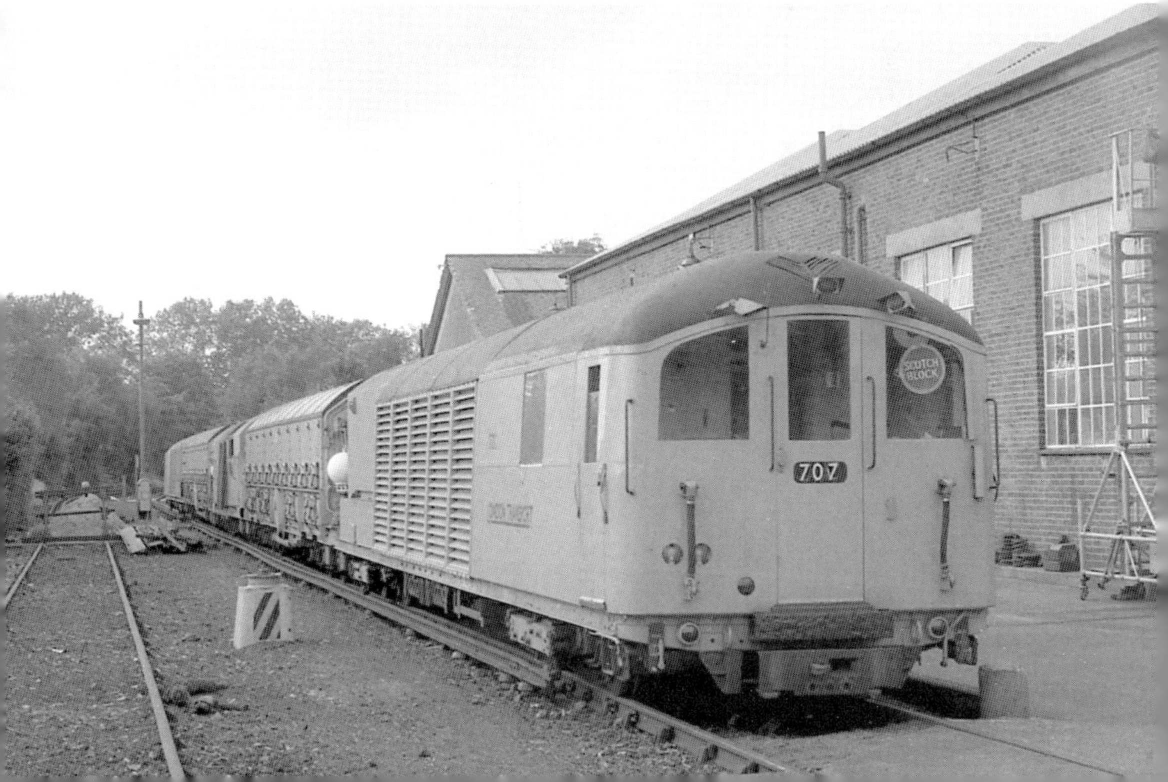
SCOTCH
BLOCK
707

Above: **Inside the** works, the body of 1962 Stock trailer 2620 is lifted off its bogies during overhaul. Beside it, non-driving motor car 9621 from the same unit will undergo similar treatment. Visible on the right is a car of recent District Line D78 Stock. These had been subject to public complaints about lack of ventilation, therefore some of their windows, which had hitherto been fixed, were altered to have passenger-opening facilities.

Opposite above: **On the** traverser at Acton Works, used to move cars about between different stages of their overhaul procedure, is Metropolitan Line A60 Stock driving motor car 5103. By this time, A Stock had already given more than twenty years' good service and would survive almost another thirty years more.

Opposite below: **The oldest** motorised resident of Acton Works on this occasion is electric sleet locomotive ESL107, one of those converted from two Central London Railway/Central Line driving motor cars, dating from 1903, joined back-to-back. It has recently been repainted in the new yellow livery, with red fleet names and numbers. Luckily, it has been saved for posterity by the London Transport Museum.

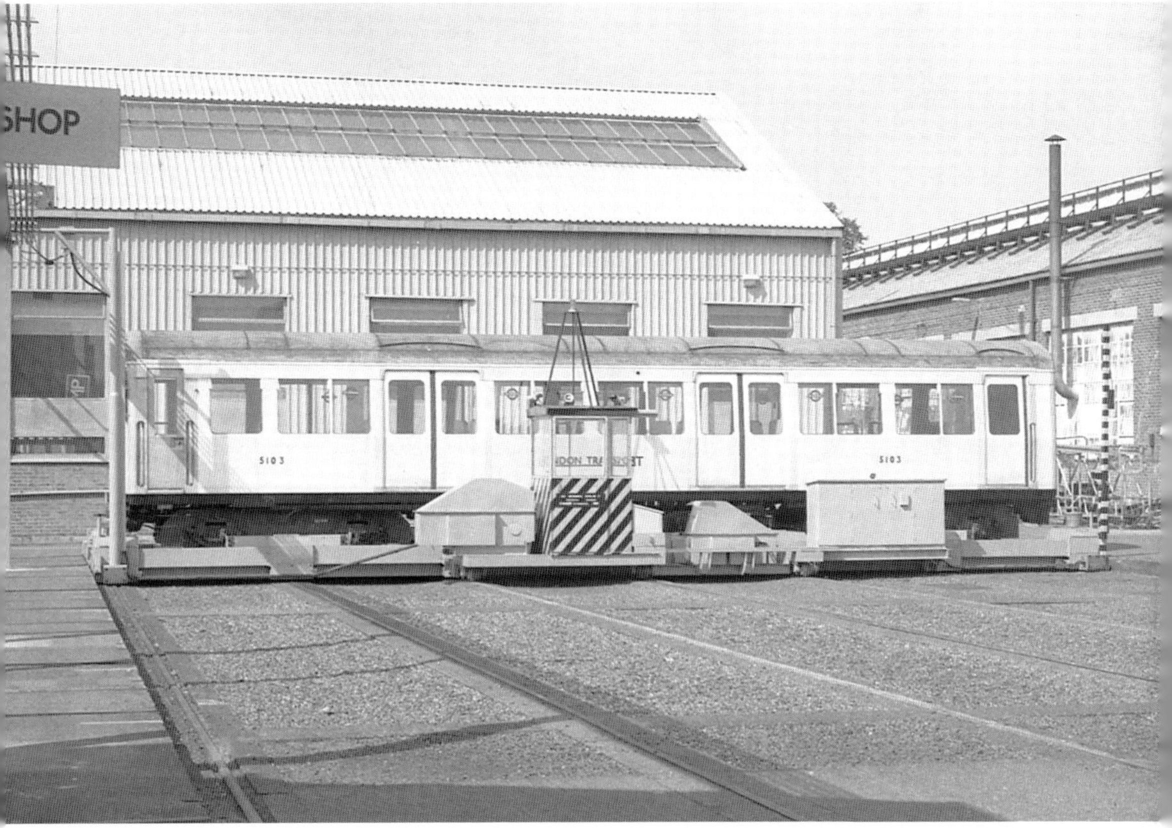
SHOP
5103
5103

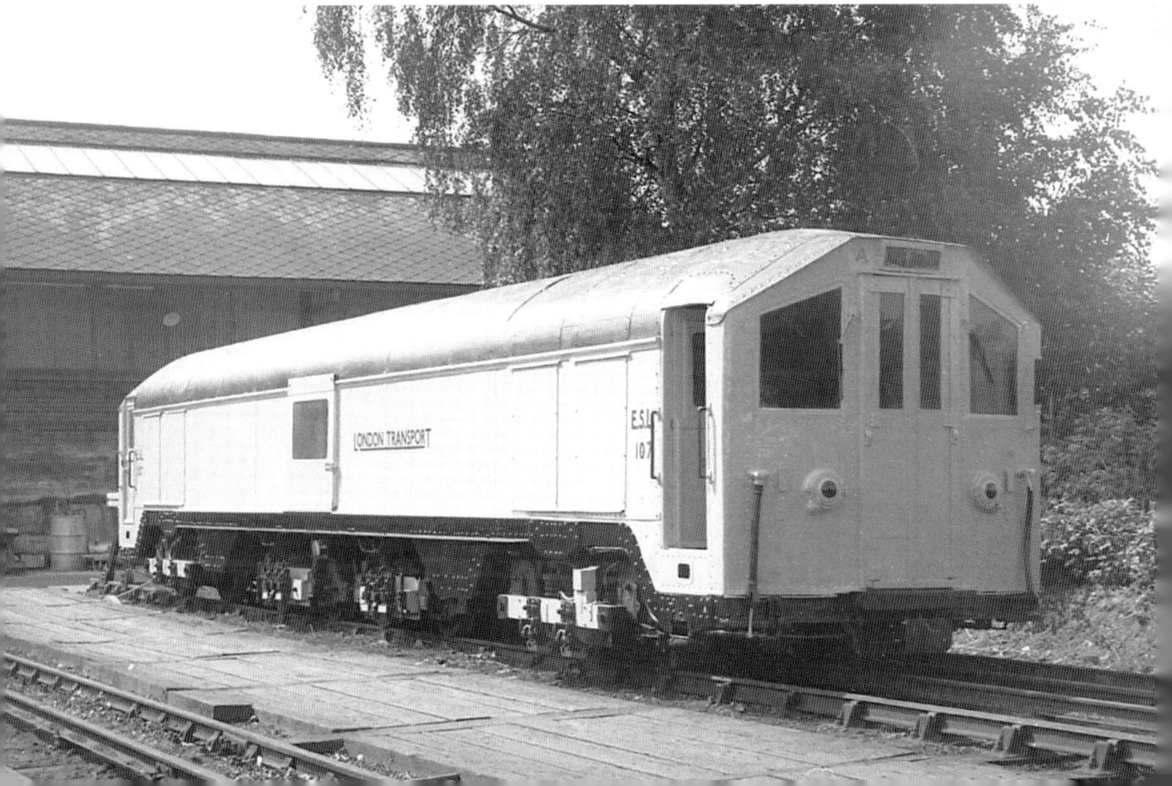
LONDON TRANSPORT

Above: **Also bearing** the recently-introduced yellow livery are pilot motor cars L126 and L127, converted from Q38 Stock driving motors 4416 and 4417. There were four such conversions in all; these two cars have also been preserved by the LT Museum, regaining their original numbers. One of the Standard Stock driving motor cars similarly converted for such duties is on the right.

Opposite above: **More than** two years after their withdrawal, a unit of CO/CP Stock remains at Acton Works on the Golden Jubilee weekend, with CP driving motor car 54264 nearest the camera. It had been used for various test purposes. Sadly, it did not survive into preservation.

Opposite below: **Resplendent in** a new coat of yellow paint after its April 1983 overhaul, Standard Stock pilot motor car L135 had been converted from 1934-built driving motor car 3701 in 1967. It has since been preserved by London Underground. Another of these cars is coupled to it in this picture.

THE LONDON UNDERGROUND, 1968–1985 • **173**

Above: **A relic** that is also now preserved today is ESL118A, along with its partner ESL118B, converted from Metropolitan Railway T Stock driving motor cars 2758 and 2749 dating from 1932. Here, they have a special truck between them used for spreading de-icing fluid when working as electric sleet locomotives in their post-passenger service careers.

Opposite above: **Back in** the land of the living, so to speak, driving motor car 1171 heads a freshly overhauled three-car set of 1959 Stock awaiting return to the Northern Line. It is of note how the roof areas and cab doors appear not as pristine as the rest of the cars' exteriors. A unit of 1938 Stock stands on the left.

Opposite below: **Battery locomotives** L51, L24 and L48 undergo attention in the Acton Works repair shop. The outer two were built for LT by British Rail in 1974, part of a batch of eleven. L24 is also one of a batch of eleven, supplied by Metro-Cammell in 1965.

THE LONDON UNDERGROUND, 1968–1985 • **175**

***Above*: L61 is** the last (numerically) of seven battery locomotives supplied by Pickering in 1951/52. Freshly overhauled in the new yellow livery, it is one of a number of London Underground exhibits on display at an open day at British Rail Eastern Region's Stratford Works on 9 July 1983. As may be observed, it is of the same general design as other battery locomotives built for LT between 1936 and 1974.

***Opposite above*: Caught by** evening sunshine, driving motor car 1252 brings a seven-car formation of 1959 Stock in to Kensal Green Station on a Bakerloo Line service to Stonebridge Park on 21 July 1983. Once again, the absence of graffiti is notable, though this dreck will all too soon infest the area, making it one of the worst in London.

***Opposite below*: On the** same day as the previous picture, a seven-car train of Bakerloo Line 1938 Stock, also bound for Stonebridge Park, has just climbed up from the tube tunnels into Queen's Park Station. It is led by driving motor car 10221 which, along with partner 11221, would become cars 122 and 222 in British Rail's Isle of Wight two-car unit 002 in 1990.

Above: **At this** period, the surviving 1960 Stock units, working the Woodford to Hainault and Epping to Ongar shuttle services, were looking very down at heel. Their two former Standard Stock trailers had been replaced by one 1938 Stock trailer painted silver, thus reducing their formations to three-car. One of these, with driving motor car 3908 leading, trailer 4921 in the middle and driving motor 3909 trailing, has just come out of the sidings at Woodford to operate a trip around the lightly-used section of the Central Line to Hainault on a drizzly 1 November 1983.

Opposite above: **In total** contrast on the same dreary November day, brand new 1983 Stock driving motor car 3601 heads a three-car unit of this type on test at South Ealing Station. Since the discontinuation of District Line services on the Hounslow West branch in the early 1960s, the centre tracks on its four-track section have been used for testing new stock, being ideal to do so in view of their suitability for both tube and surface stock, and their proximity to Acton Works. Of note here are the station's Charles Holden-designed parts dating from the Piccadilly Line's westward extensions in 1932, and the original District Railway structures, notably the waiting room on the westbound platforms on the right.

Opposite below: **On the** same occasion, driving motor car 3701 is at the eastern end of the new 1983 Stock unit. Built by Metro-Cammell and having some similarity in appearance to 1973 Stock, one of which has arrived on the adjacent platform, their frontal design lacked the wrap-around windscreens and they also had single-leaf passenger doors. In many ways they were a tube version of the District Line's D78 Stock, having the same interior decor too. Only half of the intended number of cars were initially delivered, owing to the Thatcher regime's squeeze on London Transport's finances and also a decline in passenger numbers on the Jubilee Line, where they were meant to replace 1972 Mk2 Stock. When passenger numbers increased again, the remaining cars were delivered, but they were to have a ridiculously short service life, in some cases just ten years, due to boarding and alighting delays caused by their doors as well as various other operational problems.

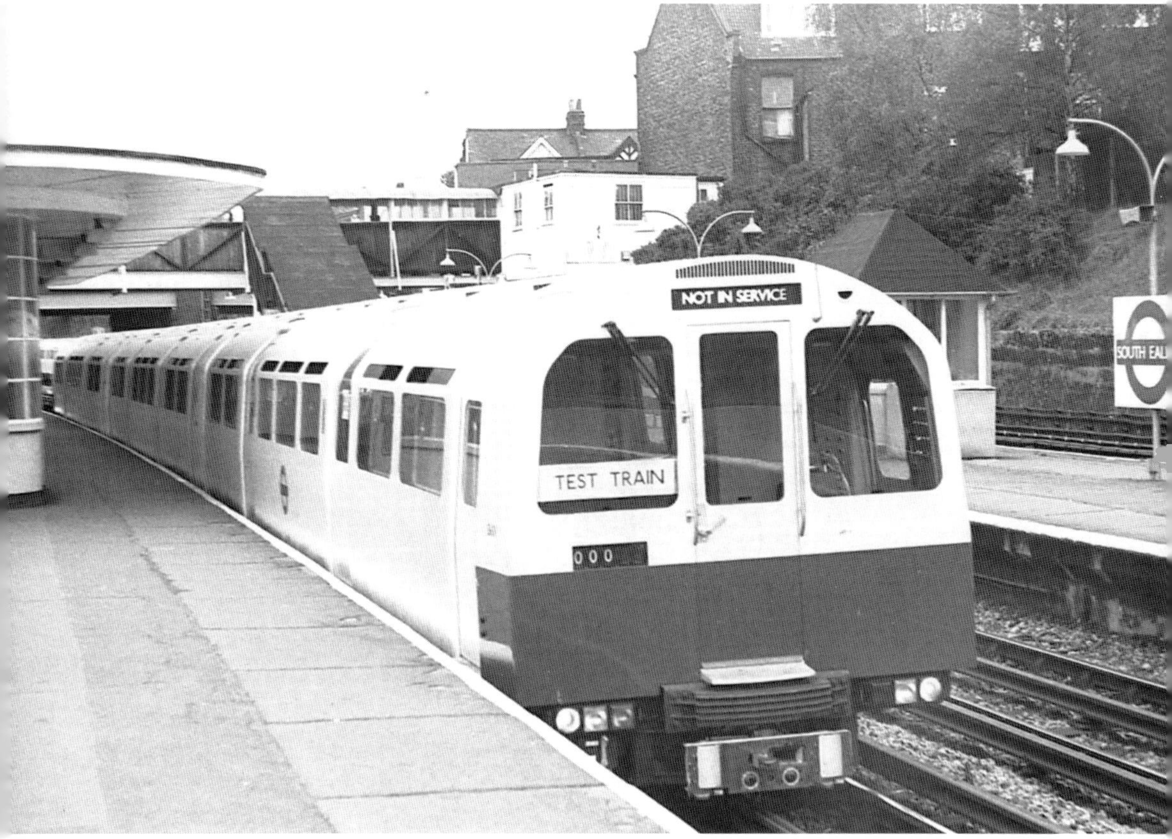
NOT IN SERVICE
TEST TRAIN
000
SOUTH EALI

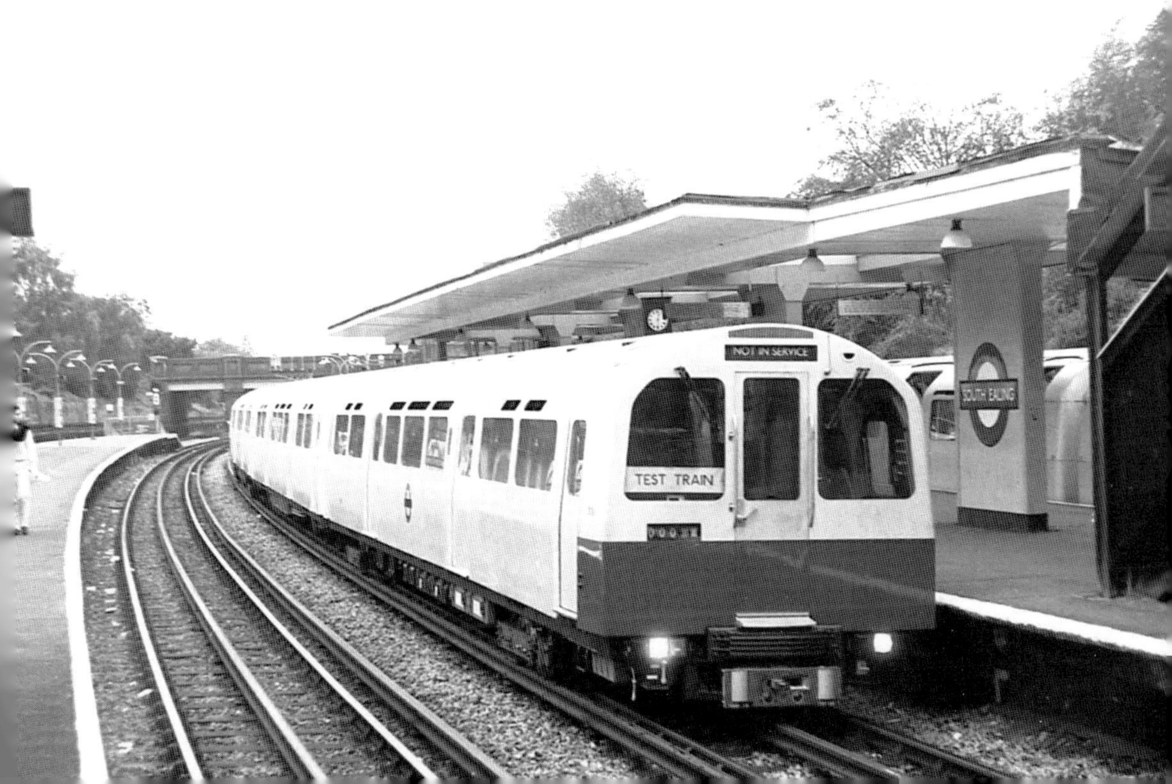
NOT IN SERVICE
TEST TRAIN
SOUTH EALING

Above: **A sad** sight on 23 March 1984, which epitomises the decline of London Underground's fortunes thanks to the Thatcher regime's polices, is the southern end of the Northern Line's Highgate Depot. Only a handful of trains are now stabled there, and tracks have been removed at its southern end, which once had extensive sidings. Highgate Wood sidings on the right have been taken out of use, sharing the fate of the branch to Alexandra Palace alongside them, which should have become part of the Northern Line in 1941, but was closed to passengers in 1954 and lifted a few years later. The two tracks on the left are on the former GNR/LNER Northern Heights branch from Finsbury Park to Highgate, East Finchley, Edgware and High Barnet, which also should have been fully incorporated into the Northern Line complex. This view is looking north from the footbridge to Park Junction signal cabin. On the left, driving motor car 1292 is heading a seven-car train of 1959 Stock facing south, whilst 1972 Mk1 Stock driving motor cars 3519 and 3218 face south inside the shed.

Opposite above: **The 1959** Stock train illustrated above has shunted right up to the mouth of Highgate west tunnel, on what was the southbound track towards Finsbury Park. Driving motor car 1195 is facing my camera at its northern end. The northbound track has been removed, and visible is the tunnel portal that was bombed during the war, and rebuilt with a round entrance, rather than the original elliptical one which the other bore still has. Much was made in a London Transport publication describing their achievements during the war of the fact that it was rebuilt very quickly, with trains running again within a couple of days. However, Underground trains *never* ran in public service through those tunnels, services at the time being LNER steam trains between Finsbury Park and Alexandra Palace or East Finchley, and of course the Northern Line's extension to Finsbury Park and Drayton Park, linking with the isolated Northern City Line, was abandoned after the war.

Opposite below: **One of** the two 1972 Mk1 trains seen earlier in Highgate Depot has just departed from its northern end to take up service for the evening rush hour on 23 March 1984. It pulls out on to what had been the northbound track from Finsbury Park. Driving motor car 3519 is at its trailing end.

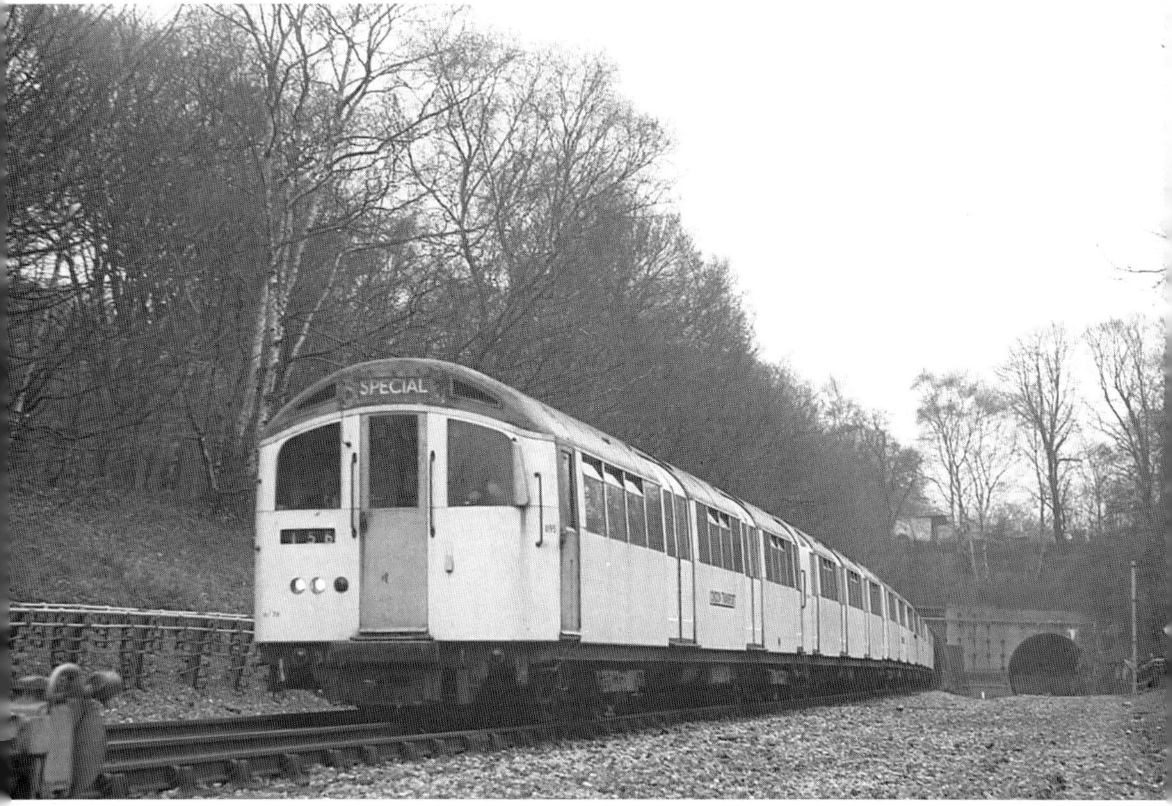
SPECIAL

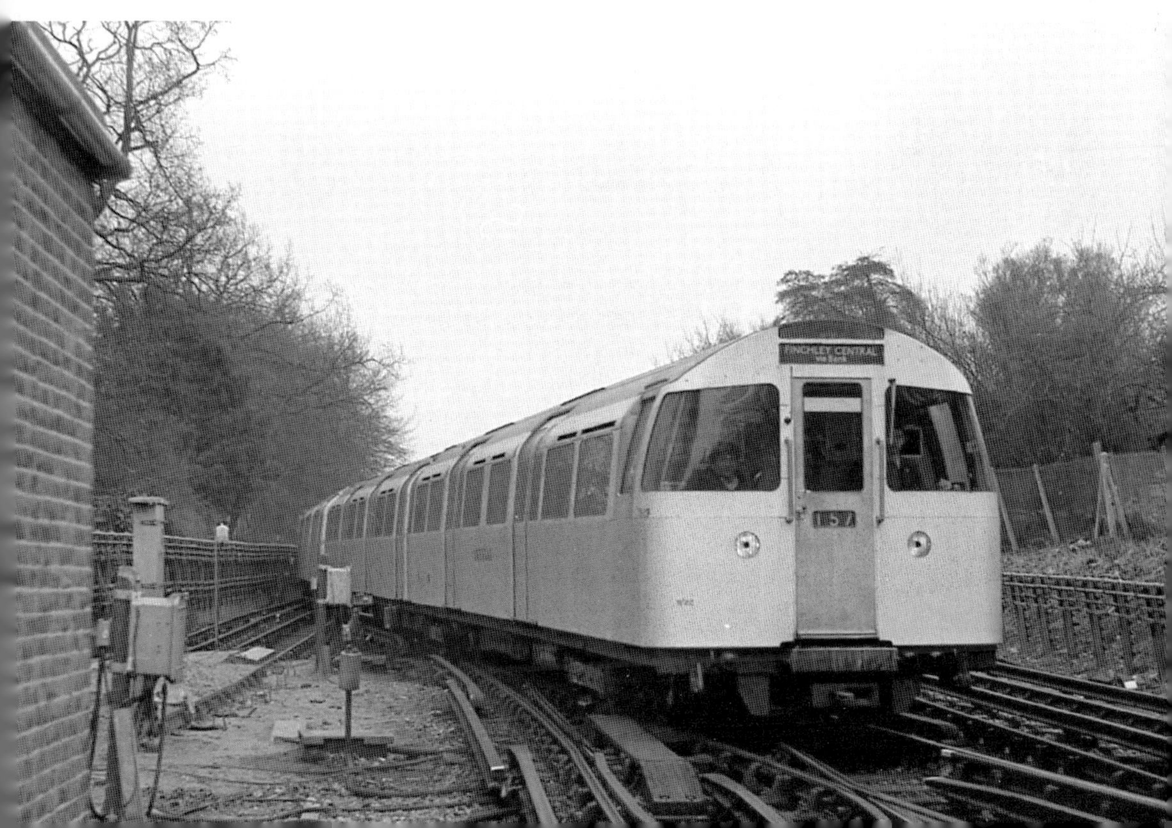
FINCHLEY CENTRAL

Above: **The same** train is held at signals on the approach to East Finchley Station, whose southern end is visible in the distance. The two tracks in this picture have only ever been used by Underground trains running in and out of Highgate Depot, and on the occasional railtour. They should have been regularly used by trains from Moorgate (Northern City Line terminus) via Finsbury Park to East Finchley and High Barnet if the Northern Line's 1935-1940 New Works Programme extensions been completed in full.

Opposite above: **When the** existing Northern Line was extended from Archway (in tube tunnels) to join and take over the LNER Northern Heights branches to Edgware and High Barnet in 1939/40, it was not possible for it to surface at Highgate and also run on to the Alexandra Palace branch too owing to the very steep gradient. Connection was therefore made just south of East Finchley Station, where the two tube tunnels emerged either side of the existing LNER tracks. Here, the train of 1959 Stock illustrated earlier at Highgate west tunnel has just taken up service at East Finchley (involving running through the station and then switching on to the southbound platform used by through trains from High Barnet and Mill Hill East) and is about to enter the tube tunnel. Driving motor car 1292 heads it. Meanwhile, the train of 1972 Mk1 stock is just entering East Finchley Station prior to taking up service. The bridge in the foreground on the left was for a public footpath from the Great North Road to Cherry Tree Wood on the right, which, of course, had to be closed when the two new outer tracks were built across it.

Opposite below: **Driving motor** car 1195 is at the trailing end of the 1959 Stock train and disappears into the Northern Line's southbound tube tunnel. It will not see daylight again until Morden; its journey there via Bank was once the longest continuous railway tunnel in Britain. The train's guard has forgotten to change the destination blind at the rear.

Above: **25 March 1984** was the last day of operation at Highgate Depot. In mid-afternoon that day, just two trains are present in this view of its northern end, with 1959 Stock driving motor cars 1112 and 1183 leading. The depot had originally been GNR/LNER carriage sheds, and only had access at its southern end. It was situated between the 'main' line from King's Cross and Finsbury Park to Edgware and High Barnet, and the branch off it to Alexandra Palace. When the extension of Northern Line services over these branches was abandoned, this proved to be an inconvenience, and the depot was subsequently rebuilt with an entrance at its northern end, facing towards East Finchley, with the sidings at its southern end, as well as Highgate Wood sidings alongside the path of the Alexandra Palace branch, later taken out of use.

Opposite above: **With no** ceremony and few photographers to record the event, driving motor car 1183 has the dubious 'honour' of heading the last train (train No.21) from Highgate depot, entering East Finchley Station at 4.20pm on 25 March 1984. The funereal, dull and drizzly weather seems to match the occasion. However, the depot was only 'mothballed' and reopened in January 1989. Subsequently, sidings to the south of the depot shed were reinstated too.

Opposite below: **On the** same occasion, another Northern Line train of 1959 Stock, headed by driving motor car 1176, calls at East Finchley Station on its way to Kennington via Charing Cross. These units were now beginning to show their age, a situation soon made much worse by the curse of moronic graffiti besmirching them.

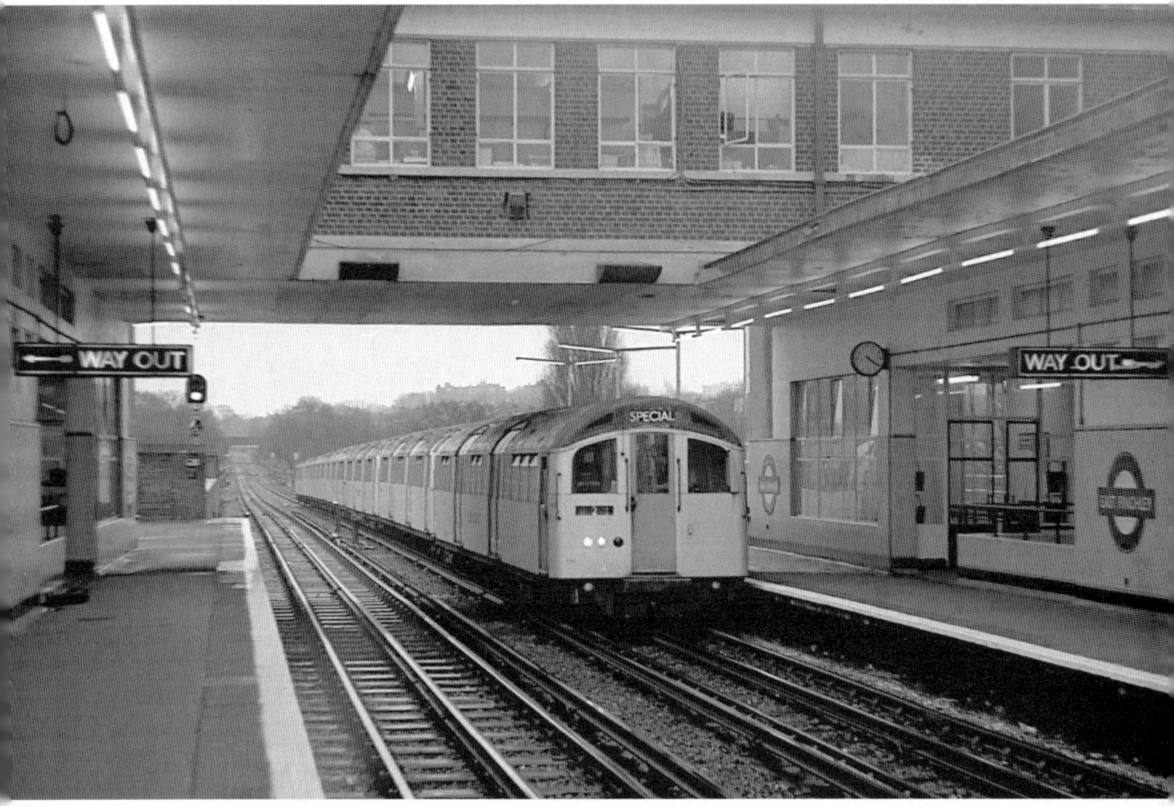
WAY OUT
SPECIAL
WAY OUT

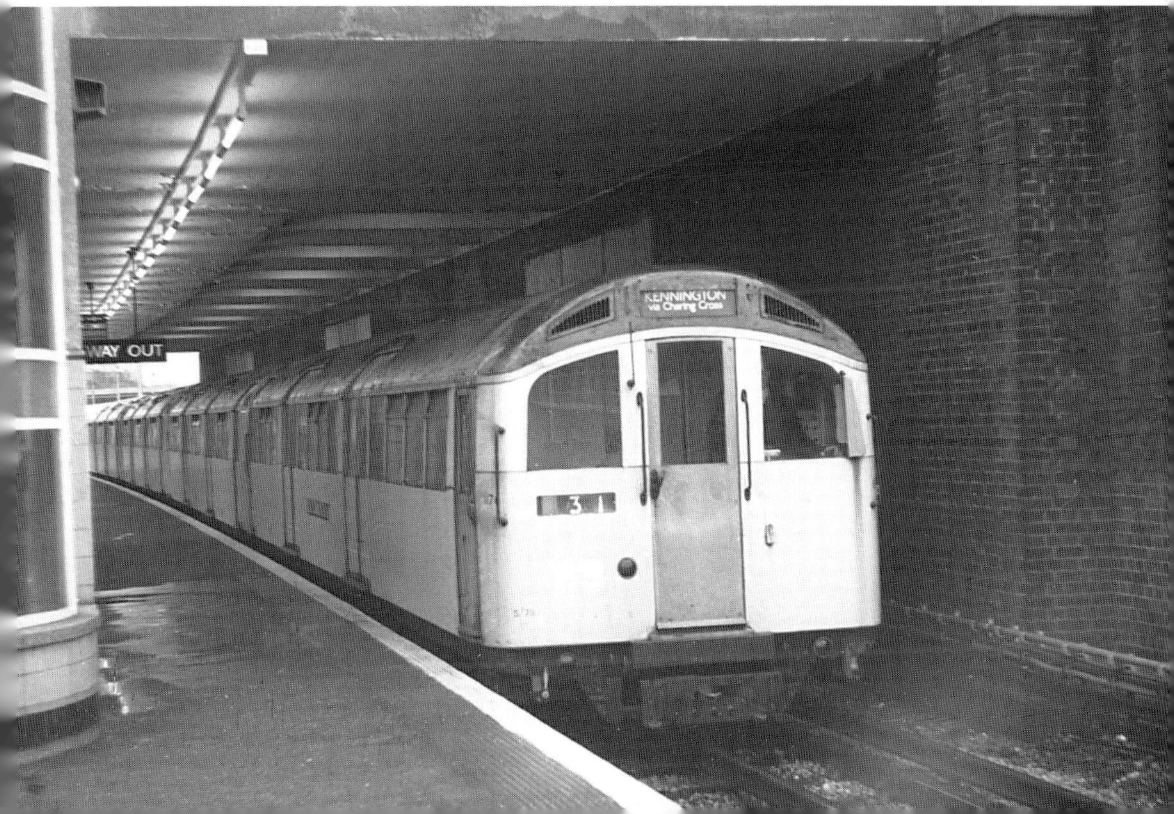
WAY OUT
KENNINGTON
via Charing Cross
3

Above: **The service** cuts that led to the closure of Highgate Depot allowed more 1959 Stock to be transferred from the Northern to the Bakerloo Line, displacing further 1938 Stock units. However, those remaining survived long enough to see services re-extended to Harrow & Wealdstone. On 21 August 1984, driving motor car 10279, one of the former LNER-owned examples, approaches Kensal Green Station on its way there.

Opposite above: **A strange** visitor to be found at Richmond Station on 15 November 1984 is G663, the gauging car converted from Standard Stock trailer 7131 in 1963. It has presumably been checking clearances on the District Line and carries the current engineering stock yellow livery. It has been brought here by 1938 Stock pilot motor cars L146 and L147, converted from cars 10034 and 11034 in 1976.

Opposite below: **On 23 November** 1984, a serious fire broke out in Oxford Circus Station, starting on the northbound Victoria Line platform. This was completely gutted, and services on the line south of Warren Street had to be suspended for just over three weeks. On 18 December 1984, the day after reopening, 1967 Stock driving motor car 3161 is at the trailing end of a northbound service in the stricken platform. As may be observed, the platform has had to be stripped back to its basic format, exposing tunnel segments and bare concrete walls. The fire began in a store used by workmen renovating the station, probably owing to a discarded cigarette end being dropped into it through a ventilator grille and igniting such things as paint stripper. The fire spread rapidly, acrid smoke being caused by burning plastic suspended ceilings. Fortunately, prompt action by London Underground staff and the London Fire Brigade saw the station quickly evacuated, and no one was killed although several people suffered from smoke inhalation. Unfortunately, London Underground did not ban smoking on their system as a result of the fire, and a much worse conflagration would follow at King's Cross almost exactly three years later.

A group of CO/CP Stock cars apparently dumped on 'Alps' sidings at the eastern end of Ealing Common Depot make a sad sight on 19 December 1984. CO Stock driving motor car 54057 is nearest the camera. A gauging device appears to be erected over one of them. They have been at the mercy of vandals who have smashed their windows, though have yet to be befouled by graffiti.

As withdrawal of the last 1938 Stock trains drew closer, one four car unit, which contained the first two such driving motor cars delivered, 10012 and 11012, was overhauled in November 1984 and given cream side window surrounds, which they had had when new. This view finds it with 11012 at its trailing end entering the gloom of Wembley Central Station on a Bakerloo Line service to Harrow & Wealdstone on 12 April 1985. The unit has subsequently been preserved as a working exhibit by the LT Museum, but the car numbered 11012 in it today is actually car 11178 renumbered.